TOPICS IN RECREATIONAL
MATHEMATICS

TOPICS IN RECREATIONAL
MATHEMATICS

BY J.H.CADWELL

CAMBRIDGE
AT THE UNIVERSITY PRESS
1970

Published by The Syndics of the Cambridge University Press
Bentley House, 200 Euston Road, London, N.W. 1
American Branch: 32 East 57th Street, New York, N.Y. 10022

Library of Congress Catalogue Card Number: 67–10013
Standard Book Number 521 04409 X

First published 1966
Reprinted 1970

First printed in Great Britain at the University Printing House, Cambridge
Reprinted by offset in Great Britain by Alden & Mowbray Ltd
at the Alden Press, Oxford

Contents

Preface

The topics discussed in this book have formed the subject of talks given by the author to mathematical and scientific audiences. The aim was to show something of the fascination and beauty of mathematics, as well as its enormous extent.

These talks convinced the author that most scientists and many mathematical specialists are unfamiliar with the problems and ideas discussed here. This fact, coupled with the belief that such material can provide great enjoyment, led him to write this book.

The mathematical background assumed does not often go beyond that of G.C.E. at Advanced Level, but it is rather more extensive than that required for many of the excellent accounts of recreational mathematics available. Like most of these accounts this book requires the persistence needed to follow a chain of reasoning of moderate length and complexity. It can be argued that the exercise of this facility is the main source of the pleasure to be found in this sort of reading. However, it is hoped that, where a formal proof has to be skipped, the reader will still be able to appreciate and enjoy the result being discussed. The various chapters are virtually independent and can be read in any order.

I would like to record my indebtedness to my colleagues Dr S. H. Hollingdale, Mr J. B. J. Thorpe and Mr D. E. Williams for their encouragement and interest in this project. In addition Dr Hollingdale has read the text and made many valuable suggestions for its improvement. I would also like to thank the Staff of the Cambridge University Press for their care and skill in planning and producing this book.

J.H.C.

Mathematics Department,
Royal Aircraft Establishment,
Farnborough

Note

Superior figures in the text refer to the lists of references at the end of each chapter.

1

Regular polyhedra

1. *The Platonic solids*

A polyhedron is a solid bounded by plane faces, the three-dimensional analogue of a plane polygon. Such a plane figure is said to be regular if all sides and angles are equal; there is evidently an infinite number of regular polygons. The best-known regular polyhedron is probably the cube, distinguished by its 6 regular polygonal faces, each of 4 sides. In addition, at each vertex 3 edges meet symmetrically. A section of the cube near a vertex, by a plane perpendicular to the diagonal through that vertex, is an equilateral triangle. We therefore describe the cube by the number pair (4, 3), a valuable symbolism due to Schläfli. The first number indicates that each face contains 4 edges, the second that 3 edges meet at each vertex.

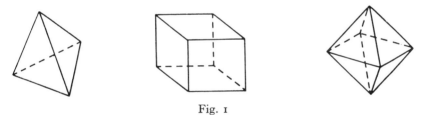

Fig. 1

In Figure 1 we have 2 other regular polyhedra. The first is a regular tetrahedron with 4 equilateral triangular faces, occurring 3 at a vertex; its Schläfli symbol is (3, 3). The third of the group is the regular octa-hedron, called (3, 4), since there are 4 faces, each of 3 sides, at any vertex.

In Figure 2 the remaining Platonic solids are depicted. The regular dodecahedron (5, 3) has 12 pentagonal faces; while the regular icosa-hedron, with 20 triangular faces, is (3, 5). Table 1 summarises this information.

In this table, and throughout the rest of this chapter, the word regular is understood to apply to any polyhedron mentioned, unless the contrary is indicated.

The first three occur in nature as crystals, and while a dodecahedral crystal with irregular pentagonal faces exists, crystallographic theory shows that neither of the last two can form a crystal and retain its regularity. All have long been known, and dodecahedral charms date from Etruscan times. A pair of icosahedral dice may be seen in one of the Egyptian rooms of the British Museum. The Greeks were much interested in these solids, hence the name Platonic. Euclid in his *Elements* discusses their properties.

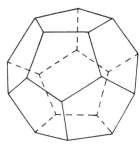 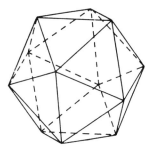

Fig. 2

Table 1. *The Platonic solids*

Name	No. of faces	No. of vertices	No. of edges	Schläfli symbol
Tetrahedron	4	4	6	(3, 3)
Hexahedron (cube)	6	8	12	(4, 3)
Octahedron	8	6	12	(3, 4)
Dodecahedron	12	20	30	(5, 3)
Icosahedron	20	12	30	(3, 5)

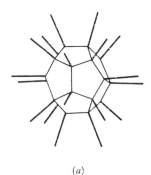 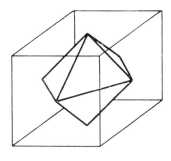

(a)   (b)

Fig. 3

Surprisingly, they occur in the skeletons of tiny sea creatures called radiolarians. These are made of silica, and measure only a fraction of a millimeter in diameter. Figure 3 (*a*) illustrates the dodecahedral version, a regular framework carrying 20 symmetrically disposed spikes. Further illustrations are given by Weyl.[6]

Figure 3 (*b*) illustrates the fact that the centres of the 6 faces of a cube are vertices of an octahedron. A similar construction applied to the octahedron yields the cube, and these solids are said to be dual to each other. The dodecahedron and icosahedron also form a dual pair, while the tetrahedron is self-dual. The Schläfli symbol (p, q) indicating p-gonal faces, arranged q at a vertex, is dual to the symbol (q, p).

2. *Proofs of completeness*

The need for proof that the above list is complete was appreciated by the mathematically mature Greek geometers, and they found such a proof. We give an alternative demonstration, based on Euler's relation for convex polyhedra. Such a solid has F faces, V vertices and E edges; these numbers are linked by the equation

$$F + V = E + 2.$$

A proof will be found in Chapter 8. Each edge belongs to 2 faces and joins 2 vertices. If there are p edges to a face, and q edges at a vertex, we count edges in 2 ways to get

$$pF = qV = 2E.$$

Combining these results with Euler's, we get

$$E = \frac{2pq}{4 - (p-2)(q-2)}.$$

It is necessary for E to be a positive integer, hence neither p nor q can exceed 5. Thus only a small number of possibilities need be examined, and just 5 are found to give a possible E. Each corresponds to one of the Platonic solids, and, within our assumptions, no others can arise.

We next outline another proof. In Figure 2 we can imagine the dodecahedron divided into 2 congruent pieces, each consisting of a pentagon bounded by 5 others. The edge of either of these cup-shaped pieces is a skew polygon of 10 sides. For the polyhedron (p, q) there is a similar skew polygon of h sides, called its Petrie polygon. Coxeter[1] shows that

$$\cos^2\frac{\pi}{p} + \cos^2\frac{\pi}{q} = \cos^2\frac{\pi}{h}.$$

The possible sets of rational values of p, q and h satisfying this relationship have been determined. They are 9 in number, 5 of them being

p	q	h
3	3	4
3	4	6
3	5	10
5/2	3	10/3
5/2	5	6

Interchange of p and q leads to the other 4. Of these sets, 5 correspond to the Platonic solids, while the other 4 involve fractional values of either p or q.

3. *The Kepler–Poinsot polyhedra*

Before dismissing the fractional solutions obtained above as meaningless, we look for analogous results with plane polygons.

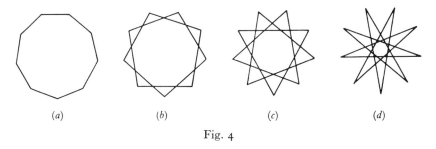

(a) (b) (c) (d)

Fig. 4

In Figure 4 the first diagram is of a simple regular nonagon or 9-sided figure. Each side subtends an angle of $\frac{2}{9}\pi$ at its centre. The second diagram can also be called regular, but each side now subtends an angle of $2\pi/(9/2)$ at the centre. We thus choose to regard it as a figure with $9/2$ sides. Figure 4(*d*) can likewise be called a regular figure with $9/4$ sides, it is easy to see that a $9/5$-sided figure is of the same form. Figure 4(*c*) consists of 3 symmetrically related equilateral triangles.

Kepler decided to include crossed polygons in the search for regularity, and discovered the first 2 solids shown in the plate (facing p. 6).

Their Schläfli symbols are $(5/2, 3)$ and $(5/2, 5)$, in each case faces being regular star pentagons of $5/2$ sides. Being unaware of the principle of duality he left the discovery of the next pair of solids to Poinsot. These have simple polygonal faces, but in each case the vertex section is a star pentagon, their symbols are $(3, 5/2)$ and $(5, 5/2)$. For each of the 4 Kepler–Poinsot polyhedra one face of the model shown in the plate has been painted in a lighter colour than the rest.

In our first proof of completeness, we disregarded the possibility of fractional p or q; indeed Euler's relation may cease to hold in such cases. The trigonometric relation of the second method depends essentially on the angle subtended at the centre of a regular polygon, and it remains valid for fractional p, q or h. Consequently it supplies a completeness proof for our list of generalized regular polyhedra.

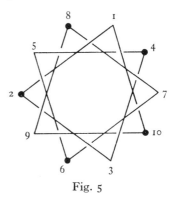

Fig. 5

The horizontal projection of the Petrie polygon for $(3, 5/2)$ appears in Figure 5. It is a regular polygon of $10/3$ sides. The Petrie polygon itself is skew, with odd-numbered vertices lying in one plane, and even numbered ones lying in a parallel plane. The side $(1, 2)$ intersects both $(4, 5)$ and $(8, 9)$ and so on.

4. *The regular compounds*

We return to Figure 4(*c*), and ask if a polyhedral analogue can be found.

The stella octangula of Figure 6, discovered by Kepler, is such an analogue. It consists of a symmetrical compound of 2 tetrahedra; they can be regarded as inscribed in a cube. Dually they circumscribe an octahedron defined by the volume common to the pair.

In Figure 7 we have a cube inscribed in a dodecahedron, and the dual situation, an octahedron circumscribing an icosahedron. The cube and the icosahedron are assumed to be opaque in the diagrams, and the visible faces of the icosahedron that lie in faces of the octahedron have been shaded. Each construction can be carried out in five ways, with the same dodecahedron and icosahedron respectively. The resulting symmetrical compounds of 5 cubes and 5 octahedra are illustrated by the

5

fifth and sixth solids of the plate. In each set, one of the 5 has been painted in a darker colour than the rest.

There is a further development: we can inscribe a stella octangula in each of the 5 cubes. This gives the seventh solid in the plate; it is a symmetrical combination of 10 tetrahedra. Finally, we can omit 5 of these to arrive at the last regular compound, one consisting of 5 tetrahedra. This can be done in two essentially different ways, only one being shown. The resulting solids, although not identical, are related as are

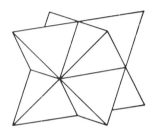

Fig. 6

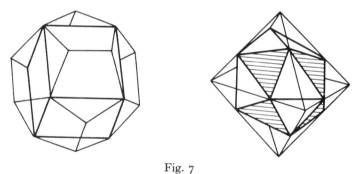

Fig. 7

object and image in a mirror, and are said to constitute an enantiomorphous pair. It has been proved that the stella octangula, together with the 4 regular compounds just discussed, exhaust all possibilities of this kind.

5. *N-dimensional space*

In Figure 8 the triangle, a two-dimensional object, is regular, but we have to put up with irregularity in our two-dimensional picture of the regular three-dimensional tetrahedron. The faces are assumed to be transparent, so all edges are drawn as full lines.

1

2

3

4

5

6

7

8

*facing p.*6

By introducing another point P, and joining it to the other 4, we obtain a two-dimensional presentation of the regular simplex in four dimensions. A three-dimensional presentation of this configuration can be constructed, but must also exhibit some irregularity. We see that the four-dimensional simplex contains 5 tetrahedra. The regular simplex in n dimensions has $(n+1)$ vertices, all pairs of which are equidistant; these pairs define its edges. It contains $(n+1)$ simplexes each of dimensions $(n-1)$.

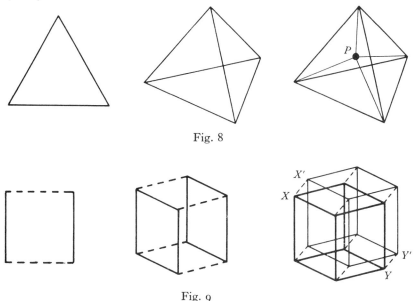

Fig. 8

Fig. 9

Figure 9 shows the measure polytopes in 2, 3 and 4 dimensions respectively. The three-dimensional version is obtained by displacing a square in a direction perpendicular to both its sides, just as a square is obtained by shifting a line in a direction perpendicular to itself. The four-dimensional version arises by moving a cube in a direction perpendicular to the 3 edges meeting at a vertex. The familiar facts that a plane can be covered with equal squares, and that 3-space can be filled with equal cubes, gives rise to the adjective measure. The word polytope describes the two-dimensional polygon, the three-dimensional polyhedron and all analogues in space of higher dimensions. We see that the four-dimensional version contains 8 cubes. These are the original cube XY, the displaced cube $X'Y'$ and the 6 cubes generated on the faces of XY.

The series of cross-polytopes is illustrated in Figure 10. In two dimensions 4 points at unit distance from the origin, and lying on a pair of perpendicular axes, are joined to form a square. In three dimensions 6 points lying on 3 mutually perpendicular lines form an octahedron. In four dimensions we can introduce a further axis PQ perpendicular to all three, and through their point of intersection. Joining P and Q to the previous 6 points the four-dimensional cross-polytope is formed. It is easy to verify that it contains altogether 16 regular tetrahedra, 8 with one vertex at P and 8 with one at Q.

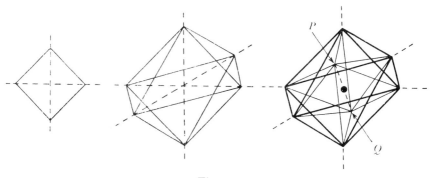

Fig. 10

We summarise the results discussed above in Table 2.

Table 2. *Structure of regular polytopes*

Type	Two-dimensional	Three-dimensional	Four-dimensional	n-dimensional
Simplex S_n	3 lines	4 triangles	5 tetrahedra	$(n+1)$ figures S_{n-1}
Measure polytope M_n	4 lines	6 squares	8 cubes	$2n$ figures M_{n-1}
Cross-polytope X_n	4 lines	8 triangles	16 tetrahedra	2^n figures S_{n-1}

We have seen how the simple Platonic solids of Figure 1 give rise to n-dimensional analogues. Other extensions will be found in Coxeter[1], where two-dimensional representations of some of them are given. Table 3 enumerates the possibilities.

There are just 3 convex cases and no stellated forms for all dimensions greater than 5. No regular compounds exist in five or six dimensions, but some are known in seven.

8

Table 3. *Numbers of regular polytopes*

Type	Two-dimensional	Three-dimensional	Four-dimensional	Five-dimensional
Convex	∞	5	6	3
Stellated	∞	4	10	0
Compound	∞	5	30	0

6. *Some further developments*

If the mid-points of edges of a cube are joined, as in Figure 11 (*a*), and the 8 corner pyramids removed, a solid known as a cuboctahedron results. It is described as a semi-regular polyhedron. It has two types of regular face, and only one type of vertex. This is no longer a regular

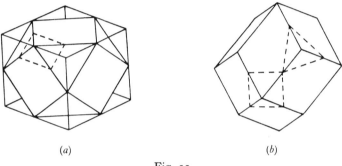

(*a*)   (*b*)

Fig. 11

vertex, for the section shown dotted in Figure 11 (*a*) is a rectangle rather than a square. This is one of the 13 Archimedean solids, discussed by Rouse Ball[4], and Cundy and Rollett[5]. If we allow stellation as well as semi-regularity there are 53 further known solids. These are beautifully drawn and photographed in Coxeter, Longuet-Higgins and Miller.[2] Had the cuts been further from the centre of the cube in Figure 11 (*a*), they would have marked out octagons on the cube faces; for one particular distance these would have been regular. The resulting facially regular solid is called the truncated cube. The truncated tetrahedron and the truncated octahedron, two other Archimedean solids, are discussed in Chapter 9.

The centres of the faces of an Archimedean solid define the dual figure. This has regular vertices of more than one type, and just one kind of non-regular face. The rhombic dodecahedron, shown in

9

Figure 11(b), is dual to the cuboctahedron, having 12 congruent rhombic faces. At some vertices 3 faces meet symmetrically, and at others 4. The dotted figures indicate equilateral triangular and square sections at these two types of vertex.

If we start with a sufficiently large block of wood, and make twelve plane cuts suitably disposed, we can form a regular dodecahedron. However, there will also be other finite solids; apart from those including faces of the original block, and potentially of infinite extent.

These pieces are illustrated in Figure 12. There are 12 pentagonal pyramids, 30 (irregular) tetrahedra and 20 triangular bi-pyramids. If the 12 pyramids are stuck to the faces of the dodecahedron we obtain the

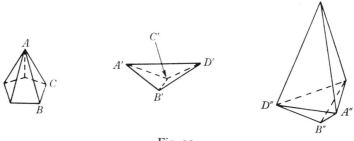

Fig. 12

solid (5/2, 5). Now the 30 tetrahedra are used to fill up gaps between pyramids. In this process the face ABC will be in contact with a face $A'B'C'$, and so on. The result is now (5, 5/2), and finally the 20 bi-pyramids are inserted in its dimples, faces like $A'B'D'$ and $A''B''D''$ being placed in contact. This gives (5/2, 3), and the process, referred to as stellation, has produced 3 of the 4 Kepler–Poinsot polyhedra.

If we generate the icosahedron in the same way, there are 472 additional pieces. From these no less than 59 solids can be formed, each having the rotational symmetry of the regular icosahedron. Very fine drawings of these will be found in Coxeter, DuVal, Flather and Petrie[3]. We have already met a few, thus (3, 5/2) and the regular compounds of 5 tetrahedra, 5 octahedra and 10 tetrahedra are all stellations of the icosahedron. Of the 59, 27 are like the compound of 5 tetrahedra, having mirror images not directly congruent to themselves.

References

1. H. S. M. Coxeter. *Regular Polytopes*, 2nd edition (Macmillan, 1963.)
2. H. S. M. Coxeter, M. S. Longuet-Higgins and J. C. P. Miller. 'Uniform Polyhedra'. *Phil. Trans.* A, **246**, 401 (1954).
3. H. S. M. Coxeter, P. DuVal, H. T. Flather and J. F. Petrie. *The Fifty-Nine Icosahedra* (University of Toronto Press, 1938).
4. W. W. Rouse Ball. *Mathematical Recreations and Essays*, 11th edition (Macmillan, 1940).
5. H. Martyn Cundy and A. P. Rollett. *Mathematical Models*, 2nd edition (Oxford University Press, 1961).
6. H. Weyl. *Symmetry* (Princeton University Press, 1952).

2

The Fibonacci sequence

1. *Defining the sequence*

Leonardo of Pisa, also known as Fibonacci (son of good-nature) was the author of a text on algebra called *Liber Abaci*. This appeared in 1202, and contained the following problem.

Rabbits are assumed to breed as follows. A pair of rabbits in its first month of life does not produce young. During the second and each ensuing month they produce a new pair. Starting with a single pair, how many pairs will there be at the beginning of months 1, 2, 3, ...? Deaths are supposed not to occur in the period considered.

The numbers sought form the Fibonacci sequence:

$$1 \quad 1 \quad 2 \quad 3 \quad 5 \quad 8 \quad 13 \quad ....$$

Each entry is the sum of the preceding pair of terms, since, at the beginning of month $(n+1)$ there will be

(i) The u_n pairs alive at the beginning of month n.

(ii) Pairs born during the nth month. These came from pairs over 1 month old, i.e. from the u_{n-1} pairs alive at the beginning of the $(n-1)$th month.

Thus we define the F-sequence by

$$u_{n+1} = u_n + u_{n-1}, \quad u_2 = u_1 = 1. \tag{1}$$

Another more plausible biological origin can be framed concerning branching habits of trees. A tree is assumed to have branches that grow indefinitely, producing no new branches during their early stages. All branches send forth a new branch at the end of their second and each succeeding year of life. The original branch $ABCDE...$ puts forth new branches at $B, C, D, E, ...$ The branch starting at B is $BC_1D_1E_1 ...$ and its first subbranch appears at D_1 and so on. At the beginning of the fifth year,

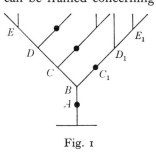

Fig. 1

12

just above the D level, there are 5 branches. Just above level E there are 8, the sixth F-number.

We shall mention other links between the F-sequence and biological growth later.

2. *Algebraic relationships*

As might be expected definition (1) leads to various algebraic results; only a few are discussed here; Coxeter [1]and Vorobyov[3] give many others. Verification of such results often depends on using (1), together with the method of mathematical induction. As an illustration we consider

$$u_{n+1}^2 - u_{n+2}u_n = (-1)^n. \tag{2}$$

Assuming this to be true for n, and using (1) with n replaced by $(n+1)$, we eliminate u_n to get

$$u_{n+1}^2 - u_{n+2}(u_{n+2} - u_{n+1}) = (-1)^n.$$

Collecting up terms, and again using (1), gives

$$u_{n+2}^2 - u_{n+1}(u_{n+1} + u_{n+2}) = (-1)^{n+1},$$

i.e. $$u_{n+2}^2 - u_{n+3}u_{n+1} = (-1)^{n+1}.$$

Thus, if the formula holds for n, it holds for $(n+1)$. It is true for $n = 1$, hence it holds for $n = 2, 3, \ldots$.

It can be rewritten in the form

$$\frac{u_{n+1}}{u_n} - \frac{u_{n+2}}{u_{n+1}} = \frac{(-1)^n}{u_n u_{n+1}}.$$

Since the right-hand side tends to 0 as n increases, we see that ratios of successive F-numbers get closer and closer to equality. This suggests that the ratio tends to a limit, a fact noted by Kepler.

We now derive a formula, due to Binet, for the nth F-number. Starting with the assumption that

$$u_n = Ax^n,$$

and substituting in (1), we get on cancelling Ax^{n-1}

$$x^2 = x + 1,$$

leading to two values for x

$$g = \frac{1+\sqrt{5}}{2} = 1.62\ldots, \quad h = \frac{1-\sqrt{5}}{2} = -0.62\ldots.$$

13

Considering the expression

$$u_n = Ag^n + Bh^n,$$

since each term on the right-hand side satisfies (1), then the whole formula must do so. The constants A and B are now chosen to make u_1 and u_2 equal to 1; the result is

$$u_n = \frac{1}{\sqrt{5}} \left\{ \left(\frac{1+\sqrt{5}}{2} \right)^n - \left(\frac{1-\sqrt{5}}{2} \right)^n \right\}.$$

As h is numerically less than 1, its nth power tends to zero as n increases, so that approximately

$$u_n \simeq \frac{1}{\sqrt{5}} g^n, \quad \frac{u_{n+1}}{u_n} \simeq g.$$

Thus the ratio of successive F-numbers tends to g; this limit is called the golden number, and we return to it later.

Finally, we prove a result discovered by Lagrange. If each term of the F-series is divided by 4, the remainders are

$$1 \quad 1 \quad 2 \quad 3 \quad 1 \quad 0 \quad 1 \quad 1 \quad 2 \quad 3 \quad 1 \quad 0 \quad 1 \quad 1 \quad \ldots .$$

This sequence is periodic, the first 6 values being repeated indefinitely. We first note that this set of remainders can be generated by (1), used in a slightly modified form. Thus

$$1+1 = 2, \quad 1+2 = 3, \quad 2+3 = 5.$$

As the third number exceeds 4, we take off 4 before proceeding. Now, on reaching the seventh and eighth terms these are seen to be the same as the first two. Consequently the process cannot fail to reproduce the same set from this point, and so on. In other words periodicity depends on the recurrence of a number pair in the sequence. The argument does not depend on the divisor used.

When dividing by k, there are just k possible remainders from 0 up to $(k-1)$. Thus there are just k^2 possible pairs of consecutive values, the first in the above sequence being (1, 1), the second (1̂, 2) and so on. In view of the finite number of possibilities a pair must eventually appear for the second time, and then the whole sequence is repeated from this duplicated pair onwards. This is an example of Dirichlet's 'pigeon-hole' principle. If more than n items are distributed among n pigeon-holes, at least one must contain 2 or more items.

We have not ruled out the possibility that, with the general divisor k

14

rather than 4, the first pair to be repeated is not the initial pair $(1, 1)$. It is left as an exercise for the reader to prove that $(1, 1)$ is always the first pair to recur.

3. Continued fractions

The golden number g satisfies

$$g = 1 + \frac{1}{g}.$$

Using this formula to replace the second g we get

$$g = 1 + \frac{1}{1 + (1/g)}.$$

Repeated use of this device leads to an infinite continued fraction for g, it is

$$g = 1 + \cfrac{1}{1 + \cfrac{1}{1 + \cfrac{1}{1 + \ldots}}}.$$

This fraction can be truncated at any point and the resulting finite fraction simplified, thus

$$1 + \cfrac{1}{1 + \cfrac{1}{1 + \cfrac{1}{1}}} = \frac{5}{3}.$$

We call $\frac{5}{3}$ a convergent of the continued fraction; the whole series of convergents is

$$\frac{1}{1} \quad \frac{2}{1} \quad \frac{3}{2} \quad \frac{5}{3} \quad \frac{8}{5} \quad \ldots.$$

This is the series of ratios of successive F-numbers, so continuation is easy. Successive convergents are alternately less than and greater than g.

Such infinite fractions have long been of importance in number theory; they have also arisen in analysis. The result

$$\tan^{-1} x = \cfrac{x}{1 + \cfrac{(x)^2}{3 - x^2 + \cfrac{(3x)^2}{5 - 3x^2 + \cfrac{(5x)^2}{7 - 5x^2 + \ldots}}}},$$

shows that they are not confined to expressions involving integers. Until recently such expressions for transcendental functions were regarded as mere curiosities. However, the advent of digital computers, with the

15

need to compute standard functions like sine, tan and log, has resulted in much work in this field.

All simple periodic C.F.'s with integer elements lead to quadratic surds; the theory is given by Olds[2]. We content ourselves with illustrating the point by

$$x = 1 + \cfrac{1}{1 + \cfrac{1}{2 + \cfrac{1}{1 + \cfrac{1}{2 + \ldots}}}}.$$

It is readily verified that this implies

$$x = 1 + \cfrac{1}{1 + \cfrac{1}{1 + x}},$$

a result leading to $x^2 = 3$, so the C.F. converges to $\sqrt{3}$. The first four convergents are

$$\frac{1}{1} \quad \frac{2}{1} \quad \frac{5}{3} \quad \frac{7}{4} \quad \frac{19}{11} \quad \frac{26}{15} \quad \ldots$$

The rule for their formation is found to be

$$5 = 1 + 2 \times 2, \quad 7 = 2 + 5, \quad 19 = 5 + 2 \times 7, \quad 26 = 7 + 19, \quad \ldots,$$

for top lines, with an exactly similar form for the denominators. This differs from the simpler rule for the g convergents only in that, at alternate stages, a multiplier 2 appears. In more general cases, such as the C.F. for tan x, there is still a very simple rule for the formation of successive convergents, a fact that accounts for their utility in digital computer applications.

4. *Some geometrical relationships*

We start with a paradox related to (2). In Figure 2 the 8×8 square is dissected and reassembled to form the 5×13 rectangle! The explanation of the discrepancy in area is that $ABCD$ is not really a straight line, but a narrow parallelogram of unit area. By starting with a 21×21 square, and noting that (2) implies

$$21 \times 21 + 1 = 13 \times 34,$$

a still more convincing figure results.

Geometrical applications to provide models of biological growth were at one time much studied. Today these ideas have lost much of their

16

appeal. Coxeter[1] gives an interesting account of the segmented pattern to be seen on a pineapple or pine cone. We content ourselves here with outlining the phenomenon of phyllotaxis or leaf growth. On an elm twig successive leaves occur at intervals of half a revolution, the twig is said to exhibit $1:2$ phyllotaxis. With beech leaves the ratio is $1:3$, while

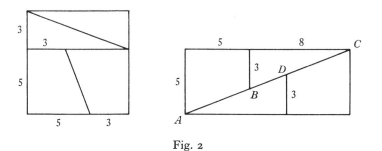

Fig. 2

with oak an advance along the stem of 5 leaves produces just 2 revolutions, i.e. the ratio is $2:5$. The theory goes on to allocate $3:8$ to poplar and $5:13$ to almond. The numbers involved are alternate values from the F-series.

5. *Golden section*

Euclid posed the question: to find a rectangle such that, when a square is cut from it, as in Figure 3, the remaining smaller rectangle has the same shape as the original. Starting with sides 1 and x, those of the reduced rectangle are $(x-1)$ and 1, so that

$$\frac{x}{1} = \frac{1}{x-1},$$

an equation whose positive root is g, the golden number or golden ratio.

In Figure 3, similar triangles can be used to show that AC, BD are perpendicular, and that $AO:OB = g:1$. Hence, the rectangle with diagonal BD is obtained from rectangle AC by rotation through a quarter-turn about O, together with a size reduction of $1:g$.

We see that BD, CE in the new rectangle have exactly the same relative positions to it as AC, BD in the old. It follows at once that BE is the diagonal of a square, and a further application proves that CF is, and so on. The points A, B, C, ... lie on a logarithmic spiral with its pole at O; Zippin[4] discusses this topic in detail.

In the past the golden ratio was made the feature of a whole mathematical theory of aesthetics. While its supposed importance in this field can be disregarded, there is no doubt at all that g appears at many places in elementary geometry.

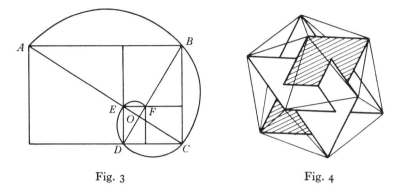

Fig. 3   Fig. 4

As an example we cite Figure 4, where three mutually perpendicular golden rectangles intersect symmetrically to form the basis of a regular icosahedron.

6. *Fibonacci sorting*

In the digital computer field it is often necessary to sort a string of numbers, stored serially on magnetic tape, into ascending or descending order. The need arises mainly in commercial applications, and often the series contains many thousand numbers. Multi-stage sorting processes are essential to deal with such large quantities of data quickly. The basic principle consists in sorting quite small blocks, perhaps of 1,000 numbers each. Then these are merged into sorted blocks of 2,000 each and so on. The merging process continues until there is a single sorted block.

Assume that there are four tape decks A, B, C and D, and that initially the unsorted data is on deck A. The first process consists of reading 1,000 numbers at a time from A into the computer. These are sorted by the machine and then the first block is copied out to deck C; the next block goes to D, and so on. Eventually half the sorted blocks will have been placed on deck C and the rest on D. The presence of an incomplete block, or of one more block on C than on D does not affect the merging process that follows.

Now the first block on C and the first on D are merged by the computer into a block of 2,000 and this is placed on A. The process works as follows. The first numbers for C and for D are read into the computer. It compares them, and assuming that ascending order is required, it copies the smaller on to deck A. If this came from C, then the second number on tape C is read, otherwise the second entry of D's block goes in. A fresh comparison is made, with the smaller number going to A, and being replaced in the same way. This process continues until one of the blocks is exhausted. The rest of the other block is then transmitted to deck A. By virtue of the selection process used, and the fact that the two blocks of 1,000 were in ascending order at the start, there is now an ascending block of 2,000 numbers on deck A.

The next block of 2,000 is merged on to B and so on. We end this stage with a number of ordered blocks of 2,000 split evenly between A and B. These now become the input blocks, and merged blocks of 4,000 are built up on C and D. Eventually this doubling process leads to a single ordered block. Fractional blocks do not interfere, since we readily see that merging two blocks does not depend on their being of the same size.

Now tape decks are expensive items, and therefore the following Fibonacci 3-deck sort has come into favour. The initial stage is designed to produce an F-number of blocks; we take 13 for illustrative purposes, but it could be much larger. They are initially stored on deck A as before; 5 of them are at once copied to B, while C is empty. The following list shows successive states of the 3 decks being used.

A	B	C
8	5	0
3	0	5
0	3	2
2	1	0
1	0	1
0	1	0

The second state arises by merging all the blocks of B with the first 5 of A, putting results on to deck C. This leaves A with 3 blocks and B free. Therefore the next merge is on to deck B, 3 blocks coming from A, and leaving it free, while the number on C is reduced to 2. The culmination of the process should now be apparent from the table

We have described the Fibonacci sorting technique in simple form, but a number of more elaborate procedures have been evolved (see Gilstad[6])

7. *Fibonacci search for a turning point*

We conclude with another recent application of the F-sequence. It is sometimes necessary to locate the maximum or minimum of a function $f(x)$ when the derivative $f'(x)$ cannot be found in analytic form. Even if the derivative can be determined, its zeros may be difficult to locate. Consequently, search methods, based on evaluating the function, have been devised. As, in some cases, the evaluation of f itself is a lengthy task, we need an optimum method of conducting the search.

Assume that the function is known to have a unique minimum at an internal point of the interval (a, b). The computing algorithm to be described determines the required value of x to within an uncertainty of $(b - a)/13$. Here 13 is the seventh F-number and 6 evaluations of $f(x)$ are needed. We assume that $a = 0$, $b = 13$; this makes description easier without leading to any loss of generality.

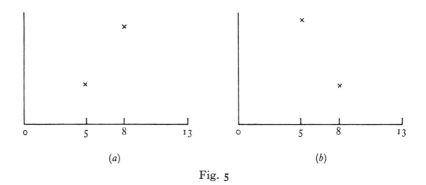

(a)   (b)

Fig. 5

The values $f(5)$ and $f(8)$ are first found, and two possibilities arise. In the first, diagram (a) of Figure 5, $f(5)$ is the smaller, and the minimum sought must lie in the range (0, 8). In case (b), $f(5)$ is the larger, and the required value of x lies in (5, 13). We see that the transformation $X = 13 - x$ turns (b) into a position completely analogous to (a). We assume that (b) holds, and describe the next step. Because of the symmetry just noted, this method, with appropriate modifications, will resolve case (a) equally well.

We next evaluate $f(10)$, and the possible results are illustrated by (a) and (b) of Figure 6. The points 5, 8, 10 and 13 have central symmetry, hence the choice of the value 10. As before, the intervals (5, 10) of (a)

20

and (8, 13) of (*b*) are symmetrical with regard to their information about $f(x)$. We note that, just as $f(5)$ exceeds $f(8)$ in (*a*), so does $f(13)$ exceed $f(10)$ in (*b*), by virtue of the initial assumptions.

Assuming that (*a*) holds, the next stages could be those shown in (*c*). The fourth evaluation is of $f(7)$, this is assumed to be greater than $f(8)$. The fifth evaluation, that of $f(9)$, giving a value less than $f(8)$, shows that the minimum lies in (8, 10). The last evaluation of $f(9 \cdot 01)$ gives the slope of f at $x = 9$ and enables us to decide between the intervals (8, 9) and (9, 10).

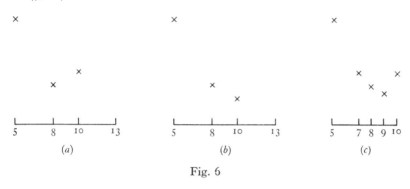

Fig. 6

By specifying more evaluations we could get greater accuracy. As the *F*-numbers are approximately multiplied by the factor g at each step, each extra evaluation divides the uncertainty of the answer by the golden number 1·62.... Further applications of this type are given by Bellman and Dreyfus[5].

References

1. H. S. M. Coxeter. *Introduction to Geometry* (Wiley, 1961).
2. C. D. Olds. *Continued Fractions* (Random House, New Mathematical Library, 1963).
3. N. N. Vorobyov. *The Fibonacci Numbers* (D. C. Heath and Co., 1963).
4. L. Zippin. *Uses of Infinity* (Random House, New Mathematical Library, 1962).
5. R. E. Bellman and S. E. Dreyfus. *Applied Dynamic Programming* (Princeton, 1962).
6. R. L. Gilstad. 'Read-Backward Polyphase Sorting', *Communications of the A.C.M.* **6**, no. 5, 220 (1963).

3

Nested polygons

1. *Sequences of polygons*

We consider a number of cyclic constructions, each of which, when applied to a polygon P of n sides, generates an n-sided polygon P'. Application of the same process to P' leads to P'', and so on. We restrict ourselves for the moment to hexagons, and denote the vertices of the initial polygon in Figure 1 (a) by the symbols A_0 to A_5.

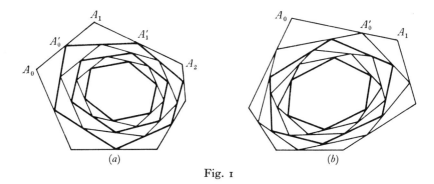

Fig. 1

In the first diagram A_0' is the mid-point of $A_0 A_1$, A_1' of $A_1 A_2$, and so on. After several stages, during which the polygons decrease steadily in size, they appear to be tending to a limiting shape. In this shape opposite pairs of sides are parallel to each other, and to the corresponding diagonals. Moreover, each polygon is approximately similar and similarly situated to the next but one in the sequence.

The trisection of $A_0 A_1$ to give A_0' and so on, leads to Figure 1 (b). The same remarks apply here, except that there now appear to be three series of similar and similarly situated hexagons.

Another linear cyclic construction consists in replacing A_1 by the mid-point of $A_0 A_2$, and so on. For convenience we refer to this as 'first diagonal bisection'. This time the limiting shape behaves very differently. In Figure 2 the first diagram shows three stages of the process, in the

22

second a further polygon has been generated from an enlargement of
the third stage. The limiting shape now appears to consist of a line
counted 6 times. Moreover, it does not shrink to zero size, as did the
polygons in previous constructions.

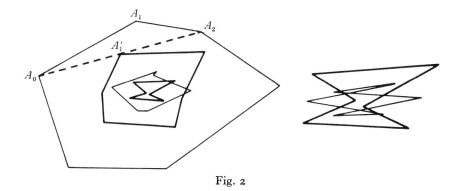

Fig. 2

In the next construction, that of Figure 3, A_0' divides $A_0 A_3$ in the ratio
$p:(1-p)$, with $p < 0.5$. Proceeding cyclically we have A_3' dividing the
same interval in the ratio $(1-p):p$. Thus the new diagonal has the same
mid-point as the old, its length being multiplied by $(1-2p)$. Repeated

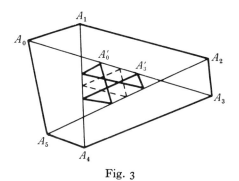

Fig. 3

application leads to a limiting triangle counted twice; it is formed by
joining the mid-points of the 3 original diagonals. This is shown dotted
in Figure 3 where $p = 0.4$. As diagonals are multiplied by 0.2 at each
stage convergence is rapid, and the second stage (not shown) is quite
close to the limit.

2. *The Fourier representation of a polygon*

Consider the hexagon with vertices A_r at (x_r, y_r) for $r = 0, 1, \ldots, 5$. The six equations

$$x_r = X + P_1 \cos \tfrac{1}{3}r\pi + Q_1 \sin \tfrac{1}{3}r\pi + P_2 \cos \tfrac{2}{3}r\pi + Q_2 \sin \tfrac{2}{3}r\pi + P_3 \cos r\pi,$$

can be solved for X, the 3 P's and the 2 Q's. By summing them, and noting that the sums of coefficients of each P and Q vanish, we get

$$6X = \sum_{r=0}^{5} x_r,$$

thus X is the x-coordinate of the centre of gravity of the 6 points. Although the results are not used, we note that a similar direct determination of each of the other constants is available, for instance

$$P_1 \sum_{r=0}^{5} \cos^2 \tfrac{1}{3}r\pi = 3P_1 = \sum_{r=0}^{5} x_r \cos \tfrac{1}{3}r\pi.$$

We next observe that, if

$$P_1 \cos \tfrac{1}{3}r\pi + Q_1 \sin \tfrac{1}{3}r\pi = C_1 \cos (\tfrac{1}{3}r\pi + \theta_1),$$

expansion of the right-hand side and comparison of coefficients leads to

$$P_1 = C_1 \cos \theta_1, \quad Q_1 = -C_1 \sin \theta_1,$$

equations easily solved for C_1 and θ_1. Thus we have a typical vertex given by

$$x_r = X + C_1 \cos (\tfrac{1}{3}r\pi + \theta_1) + C_2 \cos (\tfrac{2}{3}r\pi + \theta_2) + C_3 \cos r\pi, \qquad (1)$$

$$y_r = Y + D_1 \cos (\tfrac{1}{3}r\pi + \phi_1) + D_2 \cos (\tfrac{2}{3}r\pi + \phi_2) + D_3 \cos r\pi, \qquad (2)$$

and our analysis of the phenomena is based on this representation.

3. *The case of simple bisection*

The x-coordinates of the first derived hexagon are given by

$$x_r' = \tfrac{1}{2}x_r + \tfrac{1}{2}x_{r+1}.$$

Substituting (1) for the 2 terms on the right, and using the addition formula for cosines gives

$$x_r' = X + \tfrac{1}{2}\sqrt{3}\, C_1 \cos (\tfrac{1}{3}r\pi + \theta_1 + \tfrac{1}{6}\pi) + \tfrac{1}{2}C_2 \cos (\tfrac{2}{3}r\pi + \theta_2 + \tfrac{1}{3}\pi).$$

The effect of the process has been to multiply C_1 by $0\cdot866\ldots$, and C_2 by $0\cdot5$. The coefficient C_3 vanishes, and additions $\frac{1}{6}\pi$, $\frac{1}{3}\pi$ are made to the 2 angles. Repeated applications will cause both the trigonometric terms to diminish, but the second does so faster than the first. After l steps we neglect this smaller term to get

$$x_r = X + C \cos\left(\tfrac{1}{3}r\pi + \theta\right), \tag{3}$$

$$y_r = Y + D \cos\left(\tfrac{1}{3}r\pi + \phi\right), \tag{4}$$

where C, θ are given by

$$C = \left(\tfrac{1}{2}\sqrt{3}\right)^l C_1, \quad \theta = \theta_1 + \tfrac{1}{6}\pi l,$$

with similar expressions for D and ϕ. For large l, C and D are small, and all vertices lie close to (X, Y), i.e. the hexagon shrinks to zero size.

On expansion (3) and (4) give

$$x_r - X = P \cos\tfrac{1}{3}r\pi + Q \sin\tfrac{1}{3}r\pi, \tag{5}$$

$$y_r - Y = R \cos\tfrac{1}{3}r\pi + S \sin\tfrac{1}{3}r\pi. \tag{6}$$

We write
$$\xi_r = \cos\tfrac{1}{3}r\pi, \quad \eta_r = \sin\tfrac{1}{3}r\pi, \tag{7}$$

and consider the transformation

$$x - X = P\xi + Q\eta, \tag{8}$$

$$y - Y = R\xi + S\eta. \tag{9}$$

If the slope of the line joining points 1 and 2 in the (ξ, η) plane is μ, and for the corresponding transformed points is m, we have

$$m = \frac{y_2 - y_1}{x_2 - x_1} = \frac{R(\xi_2 - \xi_1) + S(\eta_2 - \eta_1)}{P(\xi_2 - \xi_1) + Q(\eta_2 - \eta_1)} = \frac{R + S\mu}{P + Q\mu}.$$

Since m depends only on μ, parallel lines in one system yield parallel lines in the other. If we solve (8) and (9) for (ξ, η) we see that

$$\xi^2 + \eta^2 \to L(x - X)^2 + 2M(x - X)(y - Y) + N(y - Y)^2,$$

so that the unit circle transforms into a conic in the (x, y) plane.

The six values of (7) when r goes from 0 to 5 correspond to vertices of a regular hexagon inscribed in the unit circle as in Figure 4(a). The hexagon, defined by (5) and (6), is obtained from it by the transformation (8), (9). It is therefore inscribed in a conic centre (X, Y), and has its sides parallel to its diagonals, the property noted in Section 1.

Finally, we see that the polygon coming 2 stages after (3), (4) is

$$x_r^* = X + \tfrac{3}{4}C \cos (\tfrac{1}{3}r\pi + \theta + \tfrac{1}{3}\pi),$$

$$y_r^* = Y + \tfrac{3}{4}D \cos (\tfrac{1}{3}r\pi + \phi + \tfrac{1}{3}\pi),$$

here we have introduced the factor $\tfrac{1}{2}\sqrt{3}$ and added the angle $\tfrac{1}{6}\pi$ twice in each coordinate. It follows that

$$x_r^* - X = \tfrac{3}{4}(x_{r+1} - X), \tag{10}$$

$$y_r^* - Y = \tfrac{3}{4}(y_{r+1} - Y). \tag{11}$$

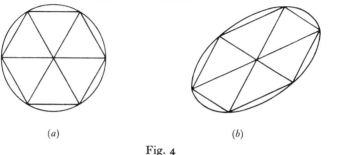

(a)   (b)

Fig. 4

Thus, apart from a renaming of the vertices, polygons occurring two stages apart are similar and similarly situated.

4. *The three-dimensional case and trisection*

Had we considered a skew hexagon in 3-space, we should have supplemented (5) and (6) by

$$z_r - Z = T \cos \tfrac{1}{3}r\pi + U \sin \tfrac{1}{3}r\pi. \tag{12}$$

Eliminating sine and cosine terms from (5), (6) and (12) gives

$$\begin{vmatrix} x_r - X & P & Q \\ y_r - Y & R & S \\ z_r - Z & T & U \end{vmatrix} = 0.$$

On expansion we get

$$l(x_r - X) + m(y_r - Y) + n(z_r - Z) = 0,$$

a result holding for $r = 0, 1, \ldots, 5$. In other words, the 6 vertices of the limiting form lie in the plane

$$l(x - X) + m(y - Y) + n(z - Z) = 0.$$

26

As this vanishes for $x = X$, etc., the plane goes through the centre of gravity of the original vertices. The limiting polygon is still plane, although arising from an initially skew figure; it will have the same properties as those found in Section 3.

If we look now at the trisection construction

$$x'_r = \tfrac{1}{3}x_r + \tfrac{2}{3}x_{r+1},$$

then (1) gives

$$x'_r = X + \tfrac{1}{3}\sqrt{7}\, C_1 \cos\left(\tfrac{1}{3}r\pi + \theta_1 + \psi_1\right)$$
$$+ (1/\sqrt{3})\, C_2 \cos\left(\tfrac{2}{3}r\pi + \theta_2 + \psi_2\right) - \tfrac{1}{3}C_3 \cos r\pi.$$

As before the C_1 term dominates, and the limiting form is

$$x_r = X + C \cos\left(\tfrac{1}{3}r\pi + \theta\right).$$

Considering the polygon three further on in sequence, we have

$$x_r^* = X + \tfrac{7}{27}\sqrt{7}\, C \cos\left(\tfrac{1}{3}r\pi + \theta + 3\psi_1\right).$$

In addition, the value of ψ_1 leads to

$$3\psi_1 = 122 \cdot 7^\circ \simeq \tfrac{2}{3}\pi,$$

and we get the approximate equality

$$x_r^* - X \simeq \tfrac{7}{27}\sqrt{7}\, (x_{r+2} - X).$$

So here the property of being similar and similarly situated is only approximately true, even in the limit. The parallelism properties of the limiting hexagon hold as for bisection.

5. *The other hexagon constructions*

Considering the first diagonal bisection of Figure 2, we have

$$x'_r = \tfrac{1}{2}x_{r-1} + \tfrac{1}{2}x_{r+1},$$

which, together with (1), gives

$$x'_r = X + \tfrac{1}{2}C_1 \cos\left(\tfrac{1}{3}r\pi + \theta_1\right) - \tfrac{1}{2}C_2 \cos\left(\tfrac{2}{3}r\pi + \theta_2\right) - C_3 \cos r\pi.$$

Here the third term is dominant, and the limiting shape is the line joining

$$(X - C_3,\ Y - D_3),\quad (X + C_3,\ Y + D_3),$$

counted 6 times, odd vertices are at one point and even vertices at the other. The multiplier in the dominant term has the numerical value unity, so that the limiting figure is now of finite size.

The diagonal division of Figure 3 leads to

$$x_r' = (1-p)x_r + px_{r+3},$$

and the application of (1) yields

$$x_r' = X + (1-2p)\,C_1 \cos\left(\tfrac{1}{3}r\pi + \theta_1\right)$$
$$+ C_2 \cos\left(\tfrac{2}{3}r\pi + \theta_2\right) + (1-2p)C_3 \cos r\pi.$$

Since $p < 0.5$, the second cosine is dominant, and as r goes from 0 to 5 we get 3 points, each counted twice. Thus the limiting form is a triangle, and again the unit multiplier of the dominant term indicates a non-zero result. With $p = 0.4$, as in Figure 3, the 3 multipliers are 0.2, 1 and 0.2; their relative magnitudes indicate rapid convergence.

6. Some results for pentagons

When considering pentagons our basic representation is

$$x_r = X + C_1 \cos\left(\tfrac{2}{5}r\pi + \theta_1\right) + C_2 \cos\left(\tfrac{4}{5}r\pi + \theta_2\right).$$

For simple bisection of sides this yields

$$x_r' = X + C_1 \cos\tfrac{1}{5}\pi \cos\left(\tfrac{2}{5}r\pi + \theta_1 + \tfrac{1}{5}\pi\right)$$
$$+ C_2 \cos\tfrac{2}{5}\pi \cos\left(\tfrac{4}{5}r\pi + \theta_2 + \tfrac{2}{5}\pi\right),$$

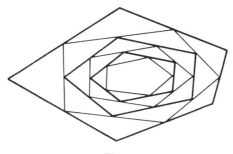

Fig. 5

and the first term dominates. The limit is a transformed regular pentagon, so that in it sides will be parallel to diagonals again. The additive $\tfrac{1}{5}\pi$ again leads to alternate pentagons being similar and similarly placed in the limit. Figure 5 illustrates these results.

Turning to first diagonal bisection we get

$$x'_r = X + C_1 \cos \tfrac{2}{5}\pi \cos \left(\tfrac{2}{5}r\pi + \theta_1\right) + C_2 \cos \tfrac{4}{5}\pi \cos \left(\tfrac{4}{5}r\pi + \theta_2\right). \quad (13)$$

Here the second term dominates, the limiting shape being

$$x = X + C \cos \left(\tfrac{4}{5}r\pi + \theta\right).$$

As r goes from o to 5, the points of the unit circle are at angles

$$\text{o,} \quad \tfrac{4}{5}\pi, \quad \tfrac{8}{5}\pi, \quad \tfrac{12}{5}\pi, \quad \tfrac{16}{5}\pi.$$

They thus form a regular star pentagon, and the limiting shape is crossed. In Figure 6 the first diagram shows two stages of the construction. The result of the second stage is then enlarged and a third polygon, this time a fully crossed pentagon, derived from it.

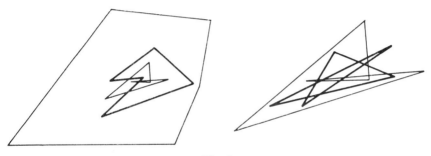

Fig. 6

However, if this construction is applied to a regular convex pentagon, the resulting sequence is one of regular convex figures. The apparent conflict with the above result arises since a regular set of points leads to a zero value of C_2 in (13). Hence C_1 now dominates and the limit is a convex figure.

7. The general theory in matrix form

For an n-sided polygon we consider the cyclic generating relation

$$x_1^{(1)} = \alpha_1 x_1 + \alpha_2 x_2 + \ldots + \alpha_n x_n,$$

with the sum of coefficients α equal to 1. The n results can be summarised by

$$\tilde{x}^{(1)} = A\tilde{x},$$

where

$$A = \begin{pmatrix} \alpha_1 & \alpha_2 & \cdots & \alpha_n \\ \alpha_n & \alpha_1 & \cdots & \alpha_{n-1} \\ \vdots & \vdots & & \vdots \\ \alpha_2 & \alpha_3 & \cdots & \alpha_1 \end{pmatrix} \quad \text{and} \quad \tilde{x} = \begin{pmatrix} x_1 \\ x_2 \\ \vdots \\ x_n \end{pmatrix}.$$

When the process is applied k times

$$\tilde{x}^{(k)} = A^k \tilde{x}.$$

The matrix A has n latent roots λ_j, and corresponding vectors $\tilde{u}_j$, such that

$$A\tilde{u}_j = \lambda_j \tilde{u}_j. \tag{14}$$

The vector $\tilde{x}$ can be expressed as follows:

$$\tilde{x} = c_1 \tilde{u}_1 + c_2 \tilde{u}_2 + \ldots + c_n \tilde{u}_n.$$

Hence, by virtue of (14)

$$\tilde{x}^{(1)} = c_1 \lambda_1 \tilde{u}_1 + c_2 \lambda_2 \tilde{u}_2 + \ldots + c_n \lambda_n \tilde{u}_n.$$

If applied k times we have

$$\tilde{x}^{(k)} = c_1 \lambda_1^k \tilde{u}_1 + c_2 \lambda_2^k \tilde{u}_2 + \ldots + c_n \lambda_n^k \tilde{u}_n,$$

and the root λ of greatest numerical value will determine the limiting vector.

There is always a dominant root $\lambda = 1$, with a corresponding latent vector of units. This indicates that, if no other root of modulus as big as 1 occurs, the figure converges to the centre of gravity of the initial set of points. Other roots occur in complex pairs, with a second real root if n is even. Let the pair of largest modulus be

$$\lambda_m \exp(\pm i\phi_m).$$

It can be proved that the complex roots are given by

$$\lambda_t \exp(i\phi_t) = \sum_{j=0}^{(n-1)} \alpha_{j+1} \exp\left(\frac{2\pi jit}{n}\right) \quad (t = 1, 2, \ldots, [\tfrac{1}{2}(n-1)]),$$

so that λ_m is easily determined. The corresponding latent vectors are of the form

$$(1, \omega^{\pm t}, \omega^{\pm 2t}, \ldots, \omega^{\pm(n-1)t}),$$

where ω is a primitive nth root of unity.

30

We leave the matrix theory at this point, but the results will be similar to those obtained earlier by Fourier techniques, terms like

$$\cos \frac{2r\pi}{n} \quad \text{and} \quad \sin \frac{2r\pi}{n}$$

entering via the complex roots of unity.

Reference

J. H. Cadwell. 'A Property of Linear Cyclic Transformations'. *Math. Gaz.* **37**, no. 320, p. 85 (1953).

4

The distribution of prime numbers

I. *The sieve of Eratosthenes*

In order to list the prime numbers less than N we can start by writing out a list of the integers from 2 to N. We ring 2 and cross out all its multiples. The next number 3 is uncrossed, so it is also ringed and its multiples crossed out. Now 4 has been deleted, so 5 is ringed and its multiples removed, and so on. We note that the process of examination need not go beyond M, the square root of N. For any composite number less than N must necessarily have at least one prime factor less than M, and it will have been deleted by the time we reach M. The Greeks regarded this process as a means of listing the primes up to N; Meissel showed that it can be used to count them, provided we have a list up to M.

We define $\pi(N)$ as the number of primes less than or equal to N; it is a quantity of central importance in our topic. For instance,

$$\pi(1,000) = 168,$$

i.e. there are 168 primes among the first 1,000 integers. Meissel's formula is

$$\pi(N) = N - 1 + \pi(\sqrt{N}) - \Sigma \left[\frac{N}{p}\right] + \Sigma \left[\frac{N}{pp'}\right] - \Sigma \left[\frac{N}{pp'p''}\right] + \dots.$$

The square brackets indicate that the integer part of the quantity inside is to be taken. In the first sum all primes p less than M occur, in the second all pairs of different primes p and p', each less than M, are taken, and so on. Signs alternate from sums with an odd number of divisors to sums with an even number, and eventually a point is reached where all further sums vanish because contents of the square brackets are less than 1.

Our proof depends on the sieve of Eratosthenes. From the original $(N-1)$ numbers we subtract the number of numbers deleted because they were multiples of 2, 3, 5, …. As the ringed prime numbers themselves were not deleted, and there are $[N/p]$ multiples of p, this number is

$$\Sigma\left\{\left[\frac{N}{p}\right]-1\right\} = \Sigma\left[\frac{N}{p}\right]-\pi(\sqrt{N}),$$

since the 1 occurs for each prime up to M.

However, numbers twice deleted have been included twice in this formula. To compensate we must add back their number, which is

$$\Sigma\left[\frac{N}{pp'}\right].$$

Still further corrections are needed for numbers deleted 3 or more times. Consider a number like 360, divisible by 2, 3 and 5. It is counted 3 times in the first sum, and 3 times in the second, on account of factors 2×3, 2×5 and 3×5. Thus no allowance for its deletion has been made with the first 2 sums, so we subtract the term

$$\Sigma\left[\frac{N}{pp'p''}\right].$$

Similar corrections for numbers with 4 and more different prime factors are required. A number with k different prime factors is counted

$$\binom{k}{1}-\binom{k}{2}+\binom{k}{3}-\ldots+(-1)^{k-1}\binom{k}{k} = 1-(1-1)^k = 1$$

times, as it should be.

Meissel, using a number of ingenious devices to reduce the labour involved, calculated $\pi(10^8)$, and later Bertelsen found $\pi(10^9)$.

We conclude this section by mentioning the complete tabulation of primes less than 10 million by D. N. Lehmer. Kulik has extended these tables up to 100 million, although his results contain mistakes. Kraitchik found all the prime numbers in the range

$$10^{12}-10^4 \quad \text{to} \quad 10^{12}+10^4. \tag{1}$$

We shall see later how this fragmentary tabulation has enabled results in the theory to be tested.

2. *Some elementary results*

Euclid posed the question, are there an infinite number of primes? His proof that there are runs as follows. Suppose that their number is finite, then there will be a largest prime p_n. We construct the larger number

$$2.3.5.\ldots.p_n+1.$$

If divided by any prime up to p_n there is a remainder of 1, therefore its smallest prime factor exceeds p_n, or it is itself a prime. In either case we have constructed a prime number greater than p_n, and the only way out of this contradiction is to abandon the hypothesis that a largest prime exists.

We can write this result in the form

$$p_{n+1} \leqslant 2 \cdot 3 \cdot 5 \cdot \ldots \cdot p_n + 1. \tag{2}$$

A little experimentation shows that the upper bound it provides is extravagantly large. Rademacher and Toeplitz[4] give a strictly elementary proof for the inequality

$$p_{n+1} < \sqrt[7]{(2 \cdot 3 \cdot \ldots \cdot p_n)},$$

valid for n greater than some fixed value.

We modify Euclid's proof to show that there are infinitely many primes of the form $(4m+3)$. We first remark that the product of 2 numbers of the form $(4m+1)$ is of the same form, for

$$(4m+1)(4m'+1) = 4(4mm'+m+m')+1.$$

Next we consider the number

$$2^2 \cdot 5 \cdot \ldots \cdot p_n + 3.$$

This is of the form $(4m+3)$, and two cases arise; it may be a prime, or it may possess prime factors. These must be of the form $(4m+1)$ or $(4m+3)$ with m non-zero. Moreover, there must be at least one of the latter, since a product of factors, all of form $(4m+1)$, is also of this form. Thus, just as in Euclid's proof, we have discovered a prime $(4m+3)$ exceeding p_n. Dirichlet proved that the arithmetic progression $(am+b)$ contains an infinity of primes, unless a and b have a common factor, when all its values are composite.

We look now at the spacing of primes. If one excludes the pair 2, 3, the gap between successive primes must be an even number, otherwise one of them would be even. The smallest possible gap of 2 is suspected to occur infinitely often; the successive primes

$$10^{15} + 149341, \quad 10^{15} + 149343$$

are a very large pair of this form. So far the result has defied proof or disproof; we return to it later.

We next ask if there is an upper limit to the gap between successive primes. We can construct the set of $(n-1)$ consecutive integers

$$n!+2, \quad n!+3, \quad \ldots, \quad n!+n.$$

It is easy to see that each number is composite, so that the nearest primes to right and left of the block differ by at least n. As n is arbitrary we can therefore make this gap as large as we please.

We mention one further result, a conjecture of Bertrand's proved by Chebyshev. The theorem states that the interval $(l, 2l)$ contains at least one prime. This can be expressed in the form

$$p_{n+1} < 2p_n, \tag{3}$$

a much improved version of (2).

3. *Numerical evidence on the density of primes*

The fact that, by going a long way on in the series of integers, we can be sure of getting a very large gap between successive primes, indicates a decrease in their density as we proceed. Table 1 illustrates the situation.

Table 1. *Counts by blocks of* 100

	0–99	100–199	200–299	300–399	400–499
0	25	21	16	16	17
10^6	6	10	8	8	7
10^{12}	4	6	2	4	2

The first line gives the numbers of primes in the first 5 blocks of 100 numbers. The second gives the corresponding values from 1,000,000 to 1,000,499, and so on. There is a slow fall in density, coupled with an appreciable degree of irregularity. The last line suggests an average gap of about 28 between successive primes. We know that this gap is sometimes as small as 2 in the region of 10^{12}, so the fluctuations are to be expected.

In view of these results it would be optimistic to expect an exact formula for $\pi(N)$ to be available. However, some description of average behaviour is still possible, as we see in the following section.

4. *The prime number theorem*

Meissel suggested that $\pi(N)$ is close to

$$\frac{N}{\log_e N},$$

for large N. Gauss arrived at the same conjecture at the age of 16, later replacing it by the better approximation

$$\pi(N) \sim \int_2^N \frac{dx}{\log x}. \tag{4}$$

Table 2 illustrates just how this formula compares with exact counts. The value of $\pi(10^9)$ was deduced from the Meissel formula as explained earlier.

Table 2. *Counts of primes*

N	$\pi(N)$	Log. integral (4)
10^3	168	178
10^6	78,498	78,628
10^9	50,847,534	50,849,235

It will be seen that the proportional error falls as N increases, and the approximation gives too large a value.

Chebyshev made a serious but unsuccessful attempt to prove the asymptotic equivalence of $\pi(N)$ and the logarithmic integral, and in this work he was led to proving Bertrand's conjecture. It was not until 1896 that a proof appeared, when independent lines of approach by Hadamard and de la Valleé Poussin were successful. Both proofs involved the Riemann ζ function, and the theory of functions of a complex variable. The basis of the method is an identity of Euler's

$$\zeta(s) = \sum_{n=1}^{\infty} \frac{1}{n^s} = \Pi \left(1 - \frac{1}{p^s}\right)^{-1},$$

the product being taken over all primes p. Littlewood later showed that the difference between $\pi(N)$ and its approximation (4) changes sign infinitely often.

The theorem can be restated in other ways. Thus it implies that the interval $(N, N+\Delta N)$ contains about

$$\frac{\Delta N}{\log N}$$

prime numbers. Again this is equivalent to saying that, near the large number N, the average gap between primes is $\log(N)$.

5. A heuristic proof of the theorem

Since the original proofs, others of a more elementary nature have appeared, in particular the use of complex variables has been avoided. However, none of the known proofs are at all easy. We present an interesting statistical approach due to the Nobel prizewinner Gustav Hertz, discussed by Courant and Robbins[2]. We assume the existence of a density function $W(x)$ such that there are $W(x)\,dx$ primes between x and $(x+dx)$. This assumption is the sole reason why the proof that follows cannot be made rigorous. It alone of the various steps taken cannot be justified formally.

We shall use Stirling's asymptotic formula

$$n! \sim n^n \sqrt{(2\pi n)}\, e^{-n}.$$

From this we can obtain the simpler, but less precise, result

$$\log n! \sim n \log n, \tag{5}$$

the error involved being of order n. Next we seek to factorise $n!$, noting that it contains the prime factor p

$$\left[\frac{n}{p}\right] + \left[\frac{n}{p^2}\right] + \left[\frac{n}{p^3}\right] + \dots \tag{6}$$

times. For example $50!$ is a very large number ending with a string of 12 zeros. It contains the factor 5 once in each of the 10 numbers $5, 10, \dots, 50$; in addition 25 and 50 contain an extra 5. There are

$$25 + 12 + 6 + 3 + 1,$$

occurrences of 2, so each occurrence of 5 leads to a zero.

Neglecting the fractional differences involved, we can replace (6) by

$$\frac{n}{p} + \frac{n}{p^2} + \frac{n}{p^3} + \dots = \frac{n}{p-1} \simeq \frac{n}{p}.$$

This leads to the approximation

$$n! \simeq \Pi p^{n/p},$$

the product being taken over all primes less than or equal to n.

Taking logs and using (5) we get

$$\log n! \simeq n \sum_{p \leqslant n} \frac{\log p}{p} \simeq n \log n,$$

thus

$$\log n \simeq \sum_{p \leqslant n} \frac{\log p}{p}. \tag{7}$$

The range from o to n can be divided into intervals Δx, large themselves, but much smaller than n. An interval from x to $(x+\Delta x)$ contains $W(x)\Delta x$ primes of average size $(x+\text{o·}5\Delta x)$. Ignoring the relatively small second term in brackets, the primes in this interval contribute about

$$W(x)\frac{\log x}{x}\Delta x,$$

to the sum in (7). We conclude that

$$\log n \simeq \sum \frac{W(x) \log x \, \Delta x}{x} \simeq \int_1^n \frac{W(x) \log x}{x} \, dx,$$

on replacing the sum by an integral. However, we know that

$$\log n = \int_1^n \frac{dx}{x},$$

and comparison of the two integrals leads to the conclusion

$$W(x) \log x \simeq \text{I}.$$

Finally, we have $\quad \pi(N) \simeq \int_2^N W(x) \, dx \simeq \int_2^N \frac{dx}{\log x},$

the prime number theorem.

6. *The average number of factors of n*

A number n may be a prime and possess just one factor. If it is a power of 2 it has the maximum possible number of prime factors, i.e. $\log_2(n)$. We seek the average number of prime factors of n, taken over all values of n less than x. We adopt the definitions

$$n = 2^{k_1}.3^{k_2}.\dots.p_r^{k_r}, \quad \Omega(n) = k_1+k_2+\dots+k_r.$$

We next examine the statement

$$\sum_{n \leqslant x} \Omega(n) = \sum_{p \leqslant x}\left[\frac{x}{p}\right] + \sum_{p \leqslant x}\left[\frac{x}{p^2}\right] + \dots. \tag{8}$$

A particular prime p divides altogether $[x/p]$ numbers less than or equal to x. Of these $[x/p^2]$ will contain this factor twice and so on. Hence, if all the occurrences of p in numbers up to x are counted, there will be

$$\left[\frac{x}{p}\right] + \left[\frac{x}{p^2}\right] + \dots.$$

38

Summing this expression for all p less than x gives the total number of factors of all numbers up to x, as in (8).

Euler showed that the sum of the inverses of the prime numbers diverges; Mertens obtained the result

$$\sum_{p \leqslant x} \frac{1}{p} = \log \log x + C.$$

It is interesting to compare this with Euler's result for the natural numbers

$$\sum_{n \leqslant x} \frac{1}{n} \sim \log x + \gamma.$$

The double logarithm in Mertens's formula indicates a much slower divergence of the series when only primes occur. One of the few proved results about prime pairs P, $(P+2)$ is that the sum of reciprocals of all such P converges.

We now observe that

$$\sum_{p \leqslant x} \left[\frac{x}{p}\right] = x \log \log x + xC + k\pi(x),$$

where k is less than 1. This follows since dropping the square brackets on the left-hand side incurs a maximum error of 1 for each term, and there are $\pi(x)$ terms. The other sums in (8) satisfy the inequality

$$\sum_{p \leqslant x} \left[\frac{x}{p^2}\right] + \sum_{p \leqslant x} \left[\frac{x}{p^3}\right] + \ldots < \sum_{\text{All } p} \frac{x}{p^2} + \sum_{\text{All } p} \frac{x}{p^3} + \ldots,$$

and this latter expression can be evaluated giving

$$\sum_{\text{All } p} \frac{x}{p(p-1)} < \sum_{n=2}^{\infty} \frac{x}{n(n-1)} = x.$$

Thus

$$\log \log x + C + k\frac{\pi(x)}{x} < \frac{\sum_{n \leqslant x} \Omega(n)}{x} < \log \log x + C + k\frac{\pi(x)}{x} + 1.$$

Noting that $\pi(x)/x$ behaves like $1/\log x$, and that the central term in the expression is the average number of factors of numbers less than x, we see that this average is asymptotically equal to $\log \log x$. This number increases very slowly with x, numbers up to 10^7 average about 3 prime factors, those up to 10^{80} have a mean of 5. Hardy and Wright[3] treat this and associated topics in greater detail.

7. Some unproved conjectures

The Riemann hypothesis concerning the ζ function is that all its complex zeros have real part $\frac{1}{2}$. It has so far defied all attempts at proof. Its truth would lead to a number of interesting results concerning primes. Thus, as mentioned earlier, Littlewood proved that, for a large enough N, $\pi(N)$ eventually exceeds the integral approximation to it, in spite of the results of Table 2. Skewes proved that, if the Riemann hypothesis holds, this N satisfies

$$\log \log \log N \quad < \quad 79 \text{ or } N < 10^{10^{34}},$$

a very large number.

Lord Cherwell[1] used statistical methods akin to those of Section 5 to investigate the frequency of prime pairs of the form $P, (P+2)$. His result suggests that, for large N, the interval $(N, N+\Delta N)$ contains

$$\frac{1 \cdot 32 \Delta N}{(\log N)^2},$$

such pairs. In the region defined by (1) there are 36 pairs, while this formula gives the number 35. He produced related formulae for prime triplets like 5, 7, 11, and like 13, 17, 19.

Goldbach conjectured that any even number can be expressed as the sum of 2 primes, for instance, $82 = 79 + 3$. Using a generalised Riemann hypothesis, Hardy and Littlewood proved that any odd number could be expressed as the sum of 3 primes. A proof of the latter result, independent of the hypothesis about $\zeta(s)$, was given later by the Russian mathematician Vinogradov, but Goldbach's conjecture still defies proof.

Many fascinating aspects of number theory are brought to light by Hardy and Wright[3], in their discussion of other open and disproved conjectures.

References

1. Lord Cherwell. 'Note on the Distribution of the Intervals between Prime Numbers', *Quart. J. Math.* **17**, no. 65, 46 (1946).
2. R. Courant and H. Robbins. *What is Mathematics?* (Oxford, 1943).
3. G. H. Hardy and E. M. Wright. *An Introduction to the Theory of Numbers*, 4th edition (Oxford, 1960).
4. H. Rademacher and O. Toeplitz. *The Enjoyment of Mathematics* (Princeton, 1957).

5

Inversion in elementary geometry

1. *The inverse transformation*

An inversion is characterised by a centre O and a radius k. Any figure $\mathfrak{F}$ in 2 or 3 dimensions is made up of points typified by P. We join OP and on this line select a point P' so that P and P' lie on the same side of O and

$$OP.OP' = k^2. \tag{1}$$

When applied to each point of $\mathfrak{F}$ this construction will generate the figure $\mathfrak{F}'$ formed by the points like P', and we call $\mathfrak{F}'$ the inverse of $\mathfrak{F}$. The expression (1) is symmetrical with regard to P and P', so the relation between $\mathfrak{F}$ and $\mathfrak{F}'$ is mutual; each is the inverse of the other.

We now present a table showing how various elements behave with regard to inversion.

Table 1. *Pairs of inverse figures*

Line through O	Line through O
Line not through O	Circle through O
General circle	Circle
Plane through O	Plane through O
Plane not through O	Sphere through O
General sphere	Sphere

The first pair needs no comment; for the second we consider Figure 1. If OQ is perpendicular to the line PQ, then (1) leads to

$$OP.OP' = OQ.OQ' = k^2,$$

and hence $PQQ'P'$ is cyclic. Therefore angle $OP'Q'$ equals $Q'QP$, a right angle, and the locus of P' as P varies is a circle on OQ' as diameter. We note that we have also answered the question, 'How does a circle through O invert'?

In Figure 2 we consider the inverse of a circle, centre C, with regard to O lying in its plane, but not on it. We draw $P'C_1$ parallel to QC, and have

$$OP.OP' = k^2, \quad OP.OQ = OT^2.$$

41

These results, with the similar triangles OQC, $OP'C_1$, lead to

$$\frac{P'C_1}{QC} = \frac{OC_1}{OC} = \frac{OP'}{OQ} = \frac{k^2}{OT^2}.$$

The last quantity does not depend on P, so that both $P'C_1$ and OC_1 are fixed. Hence the locus of P' is a circle centre C_1. We note that C_1 is not the inverse of C, i.e. centres do not invert into each other.

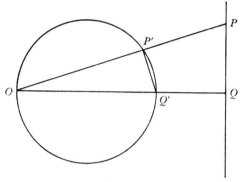

Fig. 1

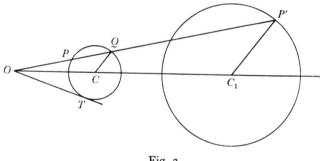

Fig. 2

The three-dimensional statements call for little comment; thus rotation of Figure 2 about OC_1 leads at once to the last result. We have proved the third if O lies in the plane of the circle. If not, the circle can be regarded as the intersection of a pair of spheres in a double infinity of ways. Since the spheres invert into spheres the inverse of their intersection is another circle.

The distinctive property of inversion is the preservation of the angle of intersection of a pair of curves. Thus, if two curves intersect at right angles, their inverses will do the same; if they touch then so do their

42

inverses. In Figure 3, a three-dimensional situation, two curves intersect at P, and their inverses cross at P'. The chords PQ, PR at P are
short in comparison with the length of OP.

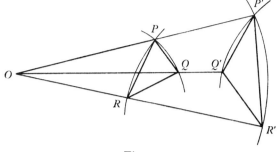

Fig. 3

By property (1) the plane quadrilateral $PP'Q'Q$ is cyclic, and similar
triangles OPQ, $OQ'P'$ lead to

$$\frac{P'Q'}{PQ} = \frac{OP'}{OQ} \simeq \frac{OP'}{OP},$$

since PQ is much smaller than OP. By symmetry $P'R'/PR$ is nearly
equal to the same ratio. A further application of this argument leads to

$$\frac{P'Q'}{PQ} \simeq \frac{P'R'}{PR} \simeq \frac{Q'R'}{QR}.$$

Hence, in the limit as Q and R approach P, the triangles PQR, $P'Q'R'$
are similar, and the angle between tangents at P equals the angle between
tangents at P'.

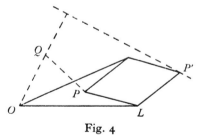

Fig. 4

Figure 4 illustrates the Peaucellier cell, a linkage designed to invert
figures mechanically. The point O is fixed, and, as P varies over a
figure $\mathfrak{F}$, P' generates the inverse figure $\mathfrak{F}'$. Proof is easy; two applications of Pythagoras lead to

$$OL^2 - (OP + \tfrac{1}{2}PP')^2 = PL^2 - (\tfrac{1}{2}PP')^2,$$

43

or
$$OP^2 + OP.PP' = OP.OP' = OL^2 - PL^2,$$

and the last quantity is fixed. The linkage will generate a straight line motion without linear guides, and was so used in the design of ventilating pumps in the Houses of Parliament in the middle of the last century.

In the diagram Q is fixed and $OQ = PQ$, so that P moves on a circle passing through O. Then P' moves along the dotted straight line. Other straight line linkages are described by Cundy and Rollett[3]. Coxeter[1] and Forder[2] also discuss inversion.

2. *Two properties of the Arbelos*

The Greeks studied the configuration of the first diagram of Figure 5 in some detail. The three semi-circles with a common diameter form a shape called the Arbelos or shoemaker's knife. We want to prove that PSR, QTR are collinear sets of points, and inversion with regard to O produces the second diagram. The circles on OP, OQ as diameters become straight lines perpendicular to $P'OQ'$, while that on PQ as diameter becomes a circle with $P'Q'$ as diameter. We note in passing

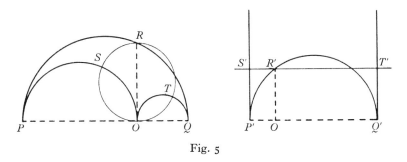

Fig. 5

that there is tangency at P' and Q', just as there was at P and Q. Further, the lines $P'S'$ and $Q'T'$, being inverses of touching semi-circles, are necessarily parallel.

The circle on diameter OR becomes a line parallel to $P'Q'$, since this circle touches PQ. Now $OP'S'R'$ is a rectangle, and therefore cyclic. The inverse of its circumcircle will be a straight line through P, S and R, so these points are collinear. The other collinearity follows in the same way.

The next property is illustrated in Figure 6. A chain of circles is

44

drawn, the nth having radius r_n and its centre at height h_n above OM. Then we have to prove that
$$h_n = 2nr_n,$$
a result obtained by Pappus without the use of inversion.

Inversion with regard to O turns semi-circles OL, OM into parallel lines, and the circle chain inverts into a new chain, all of the same

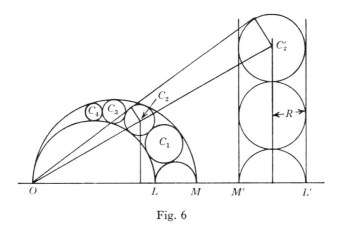

Fig. 6

radius R. The common tangent of the circle with centre C_2 and its inverse, the circle with centre C_2', has been inserted. From two pairs of similar right triangles we see that
$$\frac{r_2}{R} = \frac{OC_2}{OC_2'} \quad \text{and} \quad \frac{h_2}{4R} = \frac{OC_2}{OC_2'}.$$

This proves the result for $n = 2$, the same method applies for any value of n.

3. The problem of Apollonius

We wish to construct a circle to touch three given circles C_1, C_2 and C_3. For the moment the solution of Figure 7, with the three circles external to each other, and with the touching circle external to all three, is considered. If the smallest of the given circles is C_3, of radius r_3, we replace the problem by a new one. Circles C_1 and C_2 are replaced by concentric circles D_1 and D_2 with radii $(r_1 - r_3)$ and $(r_2 - r_3)$, while C_3 becomes the point D_3. Then the required circle, after expansion by r_3, goes through D_3 and touches D_1 and D_2.

45

Inverting with regard to D_3 this tangent circle becomes a line touching D'_1 and D'_2. Four such lines can be found; we need the direct common tangent on the opposite side of the circles to D_3. This is a standard construction, and, once the line is found, its inverse is contracted by r_3 to give the required circle.

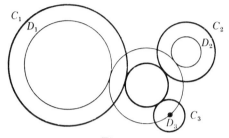

Fig. 7

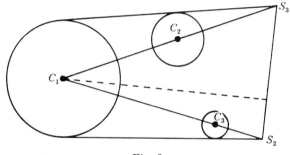

Fig. 8

There are other possibilities, the tangent circle could include within it 3, 2 or 1 of the given circles. The last two cases each arise in three ways, so that, with the solution found above, there are eight possibilities. The same method will determine each one, but now some of the given circles will expand in the initial step.

The analogous problem in three dimensions is to construct a sphere to touch four given spheres. Again one of these is reduced to a point, and inversion with regard to it generates a simpler problem. We now require the common tangent plane of three spheres. In Figure 8 C_1, C_2 and C_3 are the centres of these spheres, while S_2 and S_3, lying in their plane, arise from common tangents as shown. Any plane through S_2 touching sphere C_1 will also touch C_3; we see this by imagining the plane figure rotated about the line of centres. Similarly, a plane through S_3 touching sphere C_1 also touches C_2.

Thus we need only construct a plane through $S_2 S_3$ and touching sphere C_1. A section through the dotted line and perpendicular to the plane of the paper reduces this problem to that of drawing a tangent from a point to a circle. Thus the construction is completed; in general there will be sixteen solutions, as each of the given spheres is either inside or outside the tangent sphere.

4. *Steiner's chain of circles*

In the first diagram of Figure 9 we try to fill the region between circles C_1 and C_2 with a chain of touching circles. To see if this is possible we first choose a centre of inversion in such a way that C_1' and C_2' are concentric. The point O lies on the line of centres of C_1 and C_2 drawn below the circles, and we seek x so that the mid-points of $A_1' A_2'$ and $B_1' B_2'$ coincide. This leads to

$$\frac{k^2}{x} + \frac{k^2}{x + A_1 A_2} = \frac{k^2}{x + A_1 B_1} + \frac{k^2}{x + A_1 B_2},$$

a quadratic giving two values of x.

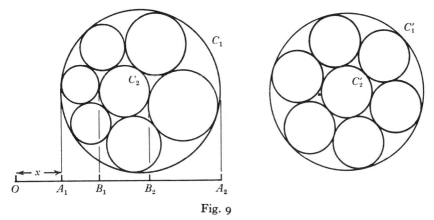

Fig. 9

In the inverse diagram we try to construct a chain of necessarily equal circles. The position of the first circle drawn is of no importance, the closure of the chain depends only on the radii of the base circles. Similarly, in the original picture, if a chain of n circles closes, then there is an infinite number of such chains. The property is said to be poristic; there is either no solution or an infinite number.

Let P_1, P_2, P be the centres and r_1, r_2, r the radii of C_1, C_2, and a circle of the chain. Then we see that

$$PP_1 = r_1 - r, PP_2 = r_2 + r,$$

so that $PP_1 + PP_2$ is the same for all circles of the chain. It follows that the centres P lie on an ellipse with foci at P_1 and P_2. I am indebted to Prof. Ogilvy[4] for this interesting result. He has extended it to the centres of the spheres of Soddy's hexlet.

5. *Soddy's hexlet of spheres*

Soddy, the discoverer of isotopes, found the following results without using inversion, but this transformation makes their proof almost trivial. We start with spheres A, B and C in contact, and seek a chain of spheres touching all 3. An arbitrary sphere S_1 is chosen to touch A, B and C; then S_2 is constructed to touch S_1, A, B and C, this was shown to be possible in Section 3. Next S_3 touches S_2, A, B and C and so on. Soddy proved that S_6, besides touching S_5, A, B and C, touches S_1, so that a chain of 6 spheres can always be found. As S_1 can be chosen in a single infinity of ways there is an infinity of such hexlets. We can visualise a chain as rotating with continuous change in sizes of its elements, but with all contact conditions always satisfied.

If we consider the case when B and C are large compared with A it is easy to envisage the situation. There will be a gap between A, B and C, and the hexlet surrounds A and threads this gap.

Inversion with regard to the point of contact of B and C greatly simplifies the problem. These spheres become a pair of parallel planes touching sphere A'. There is evidently a chain of 6 spheres equal to A', and touching it as well as the 2 planes. This chain inverts into Soddy's hexlet, it can be rotated, and so can the hexlet. Since the points of contact of the 6 equal spheres with either A' or the planes B' and C' lie on a circle, the same is true for the hexlet, i.e. its 6 points of contact with each sphere lie in a plane.

Moreover, we can find 2 spheres passing through the centre of inversion and touching the 6 equal spheres. Their centres lie on the diameter of A' perpendicular to B' and C'. It follows that the hexlet touching A, B and C is in contact with a pair of planes.

Ogilvy[4] devotes a chapter of his book to Soddy's hexlet.

6. *A theorem of Casey*

In Figure 10(*a*) the two circles of radii r_1 and r_2 intersect at A at an angle α. Applying the cosine rule to $C_1 A C_2$ gives

$$\cos(\pi - \alpha) = \frac{r_1^2 + r_2^2 - C_1 C_2^2}{2 r_1 r_2}.$$

Using Pythagoras theorem for $C_1 C_2$ we get

$$\cos \alpha = \frac{t_{12}^2 + (r_1 - r_2)^2 - r_1^2 - r_2^2}{2 r_1 r_2} = \frac{t_{12}^2}{2 r_1 r_2} - 1.$$

Since inversion does not alter α, it follows that the ratio of t_{12}^2 to $r_1 r_2$ is also invariant.

In Figure 10(*b*) the circles numbered 1 to 4 have a common tangent circle. The direct common tangent of 1 and 2 is of length t_{12} and so on. Casey's theorem states that

$$t_{12} \cdot t_{34} + t_{14} \cdot t_{23} = t_{13} \cdot t_{24}. \tag{2}$$

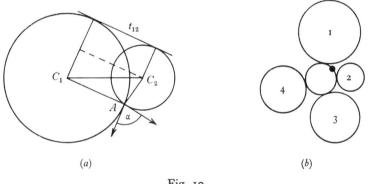

(*a*)   (*b*)

Fig. 10

We first observe that (2) is equivalent to

$$\frac{t_{12}}{\sqrt{(r_1 r_2)}} \frac{t_{34}}{\sqrt{(r_3 r_4)}} + \frac{t_{14}}{\sqrt{(r_1 r_4)}} \frac{t_{23}}{\sqrt{(r_2 r_3)}} = \frac{t_{13}}{\sqrt{(r_1 r_3)}} \frac{t_{24}}{\sqrt{(r_2 r_4)}}. \tag{3}$$

Since all the ratios are unaltered by inversion, it suffices to prove (3) in a simpler situation obtained as follows.

We invert with regard to a point on the common tangent circle lying between its contacts with 1 and 2. The inverse consists of 4 circles

49

touching a line at points A_1, A_4, A_3 and A_2 taken in that order. We note that

$$A_{12} . A_{43} + A_{14} . A_{32}$$
$$= (A_{13} + A_{32})(A_{42} - A_{32}) + A_{14} . A_{32}$$
$$= A_{13} . A_{42} + A_{32}(A_{42} + A_{14} - A_{13} - A_{32}) = A_{13} . A_{42}.$$

If this result is divided through by the square root of the product of the 4 radii of the new set of circles it gives the expression corresponding to (3) for the inverse figure. Hence (3) holds in the original situation.

7. *Coaxal circles and stereographic projection*

In Figure 11 we start with a set of concentric circles and a family of lines through their common centre C. Since any circle cuts all the lines at right angles, the same will be true of the inverse of this configuration with regard to a point O. This consists of two sets of coaxal circles; the second, coming from the family of lines, has the pair of points O' and C' in common; and these are point circles belonging to the first set.

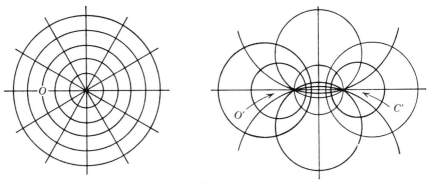

Fig. 11

This situation arises in a number of field problems in physics. An infinite conducting sheet with current entering at an electrode O' and leaving at C' has the set of circles through O' and C' as its lines of flow. The other set constitute the equipotential curves, and is cut at right angles by every line of flow.

50

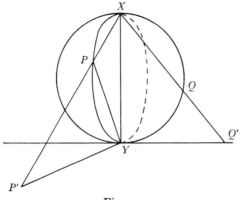

Fig. 12

In Figure 12 the sphere XY is touched by a plane $P'YQ'$ at Y, and we map the sphere on to it by projection from X. Since XPY is a right angle, it follows that

$$XP \cdot XP' = XY^2,$$

hence projection also inverts the sphere into the plane. As inversion preserves angles between curves, stereographic projection does so too. Moreover, any circle on the sphere projects into a circle in the plane.

Assuming that X is the North Pole, then lines of latitude and longitude give concentric circles and their diameters. If X is not one of the poles these lines will project into two mutually perpendicular sets of coaxal circles. Any circle of latitude cuts all circles of longitude at right angles, and this remains true of the projected figure.

References

1. H. S. M. Coxeter. *Introduction to Geometry* (Wiley, 1961).
2. H. G. Forder. *Geometry* (Hutchinson, 1960).
3. H. M. Cundy and A. P. Rollett. *Mathematical Models*, 2nd edition (Oxford, 1961).
4. C. S. Ogilvy. *Excursions in Geometry* (Oxford, 1969).

6

Some ruled surfaces

1. *The doubly ruled quadric surface*

We first note that a unique line can be drawn through a point P in 3-space to cut a pair of non-intersecting or skew lines m and n. In Figure 1 (a) join P to all points of m, thus defining a plane. Similarly, P and n define a second plane, and the two planes intersect in the required common transversal through P. In the second diagram the dotted line is to be ignored for the moment. Starting with each point of a line l we can draw through it a common transversal of the 3 skew lines l, m

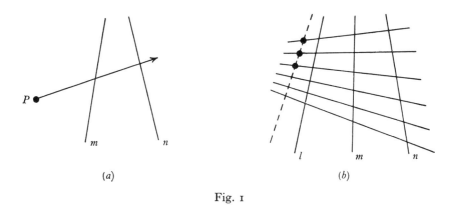

(a)   (b)

Fig. 1

and n This infinite set of lines generates a ruled surface. We now try to find the degree of this surface, i.e. the number of points in which it is cut by a general straight line a.

In Figure 2 we select an origin O on line a, and measure distances along the line from this origin. Starting with P at distance x from O we construct a point Q at distance y as follows. Through P draw the line cutting l and m. Through the point B in which it cuts l draw the line cutting n and a. Its intersection with a is Q, the point at distance y from O. Our construction of Q from P is unique, and if used in reverse yields a unique P for a given Q. It therefore seems reasonable to assume a relationship of the form

$$axy + bx + cy + d = o,$$

since this is the most general algebraic relationship in which x determines y uniquely and vice versa.

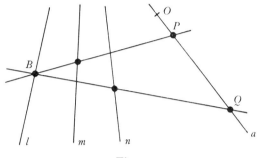

Fig. 2

The quadratic equation

$$at^2 + (b+c)t + d = o,$$

has 2 roots. If x equals either of these, the construction starting from P leads back to this point, for $y = x$ and $Q \equiv P$. This means that PB cuts n as well as l, m and a, i.e. each root of the quadratic leads to a common transversal of the 4 lines. Thus a intersects just 2 lines defining the ruled surface, and it is of degree 2; it is called a quadric surface.

We return to the second diagram of Figure 1 and insert the dotted line cutting 3 of the transversals of l, m and n. This can be done in a single infinity of ways. This dotted line cuts the quadric in 3 points, so if it had been treated as the line a above there would be 3 values of t satisfying the quadratic equation. The only way in which this can happen is for all coefficients to vanish, and then it is satisfied by all values of t. In other words the dotted line lies in the surface, cutting not just 3, but all the transversals of l, m and n. So we have 2 sets of lines, any line of one set cutting each line of the other. The section of the quadric surface by a plane is a conic. The special plane defined by one line of each system cuts the surface in just this pair of lines; this is a degenerate conic, indicating that the plane touches the quadric at the point of intersection of the lines.

We shall not prove the following result, but it is interesting to note that, if universal joints are inserted at all the intersections of lines in Figure 1(b), the surface can be deformed continuously without tearing of the joints or bending of the lines. The surface is illustrated in Figures 3(b) and 4. By squashing in the vertical direction it flattens out to a disc-like ellipse, touched by each generator. Deformations along the other axes produce plane hyperbolas; Hilbert and Cohn-Vossen[1] illustrate these changes.

2. *Quadrics in general*

The best known quadric is the ellipsoid of Figure 3(a). Its equation is

$$\frac{x^2}{a^2}+\frac{y^2}{b^2}+\frac{z^2}{c^2} = 1, \tag{1}$$

sections by planes perpendicular to any of the 3 axes being ellipses.

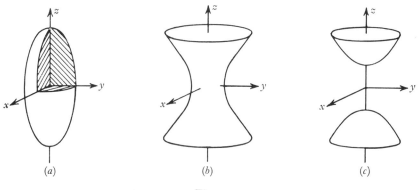

(a) (b) (c)

Fig. 3

Figure 3(b) is the ruled quadric we met above, a hyperboloid of one sheet with equation

$$\frac{x^2}{a^2}+\frac{y^2}{b^2}-\frac{z^2}{c^2} = 1. \tag{2}$$

Figure 4 shows the rulings upon this surface. We consider the pair of planes

$$\frac{x}{a}-\frac{z}{c} = \mu\left(1-\frac{y}{b}\right), \quad \frac{x}{a}+\frac{z}{c} = \frac{1}{\mu}\left(1+\frac{y}{b}\right). \tag{3}$$

For given μ they define a straight line, and as μ varies a family of such lines results. The 'product' of the two equations (3) leads to (2), so that a point satisfying both of them lies on the surface. Thus (3) gives one family of generating lines. By interchanging the signs in the 2 right-hand side terms the second family of lines results. If we try to obtain lines on (1) by the same method we get

$$\frac{x}{a}-\frac{iz}{c} = \mu\left(1-\frac{y}{b}\right), \quad \frac{x}{a}+\frac{iz}{c} = \frac{1}{\mu}\left(1+\frac{y}{b}\right). \tag{4}$$

Thus it is a ruled surface, but the lines are not real ones because of the complex coefficients in (4).

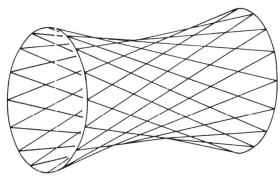

Fig. 4

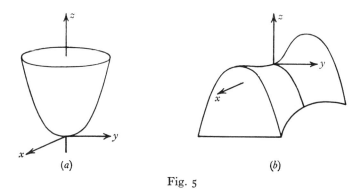

(a)   (b)

Fig. 5

The third general type of quadric with equation

$$\frac{x^2}{a^2}+\frac{y^2}{b^2}-\frac{z^2}{c^2} = -1,$$

a hyperboloid of two sheets, is also without real generators; it is illustrated in Figure 3(c).

Two special types of quadric are shown in Figure 5. The first has an equation of the form

$$z = \frac{x^2}{a^2}+\frac{y^2}{b^2}, \tag{5}$$

55

and is an elliptic paraboloid without real generators, it is shown in Figure 5(a). Figure 5(b) shows the hyperbolic paraboloid

$$z = \frac{x^2}{a^2} - \frac{y^2}{b^2},$$

a saddle-shaped surface with two sets of real generators; these are shown in Figure 6.

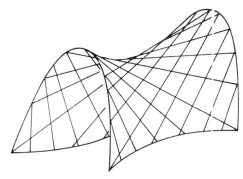

Fig. 6

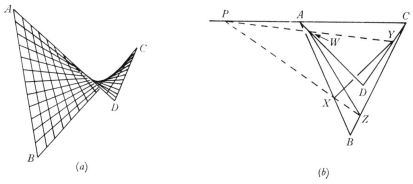

(a) (b)

Fig. 7

Figure 7(b) illustrates the construction of a hyperbolic paraboloid from a skew quadrilateral $ABCD$ with sides of equal length. One set of generators is typified by XY, where $AX = DY$. Similarly, if $CZ = DW$, then WZ gives the other set. We prove that ZW and XY intersect as follows. If ZX cuts CA at P then, by the theorem of Menelaus

$$\frac{CZ}{ZB} \cdot \frac{BX}{XA} \cdot \frac{AP}{PC} = -1.$$

Replacing lengths by equal lengths gives

$$\frac{CY}{YD} \cdot \frac{DW}{WA} \cdot \frac{AP}{PC} = -1,$$

and the converse theorem shows that P, Y and W are collinear. Thus, as ZX, YW intersect at P, they are coplanar so that XY, WZ, lying in their plane, also meet.

Hilbert and Cohn-Vossen[1] and Sommerville[2] discuss many other aspects of quadric surfaces.

3. The theorems of Pascal and Brianchon

Figure 8(a) shows a hexagon with vertices 1 to 6 inscribed in a conic. Pascal's celebrated theorem states that, if opposite sides

$$12 \text{ and } 45 \text{ meet in } X,$$
$$23 \text{ and } 56 \text{ meet in } Y,$$
$$34 \text{ and } 61 \text{ meet in } Z,$$

then these points of intersection are collinear.

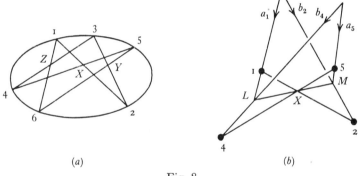

(a)

(b)

Fig. 8

Pascal's theorem can be deduced from the existence of generators on a quadric surface. We regard the conic as the intersection of a plane π and the surface. Through points 1, 3, 5 we select generators a_1, a_3, a_5 from one system; and through 2, 4, 6 pass b_2, b_4, b_6 from the other. Let

$$a_1, b_4 \text{ intersect in } L,$$
$$b_2, a_5 \text{ intersect in } M,$$
$$a_3, b_6 \text{ intersect in } N.$$

57

Then in Figure 8(b) the lines a_1, b_4, a_5, b_2 form a skew quadrilateral with one diagonal LM. It follows at once that 12 and 45 intersect on LM, i.e. X lies on LM, similarly Y, Z lie on MN, NL. As X, Y and Z also lie in plane π, they lie on its intersection with plane LMN, i.e. they are collinear.

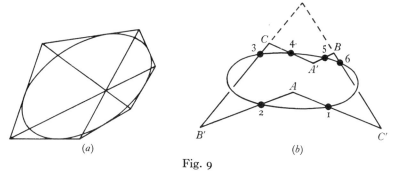

(a) (b)

Fig. 9

Brianchon's theorem states that the diagonals of a hexagon circumscribing a conic are concurrent, it is illustrated in Figure 9(a). We again consider a quadric surface and its generators. However, the relation of this quadric to the plane Brianchon configuration does not appear until the end of the argument.

The second diagram of Figure 9 shows three generators of one type through points 1, 3, 5 and 3 of the other through 2, 4, 6. The 6 points 1 to 6 again lie in a plane section π of the surface. As BC' and $B'C$ are generators of different types they intersect, and are therefore coplanar. Hence BB' and CC' meet; two similar proofs show that all three of AA', BB', CC' meet in a point.

We now assume that the tangent planes to the surface at all points of the conic in which it is cut by π meet in a point P (the pole of π). Projecting from P on to a plane, the conic yields a conic. The generator AC' lies in the tangent plane at 1, and this goes through P. Thus the projection of AC' touches the projected curve, and the six generators project into a circumscribed hexagon. The lines AA', etc., project into its diagonals which therefore concur, thus the plane projection is just Brianchon's configuration.

4. *The astigmatic surface*

Figure 10(a) illustrates the process of image formation by a lens. All rays leaving O at a given moment reach points of a spherical wave-front s

at some later time. The lens deforms this wave-front into surface t with individual rays still perpendicular to it. If t is part of a sphere of radius R, its equation will be

$$z(2R - z) = x^2 + y^2,$$

and this can be expanded in the form

$$z = \frac{x^2 + y^2}{2R} + A(x^2 + y^2)^2 + \ldots,$$

where $A = 1/8R^3$. The normals to t then pass through a true point image on the axis of the lens. In practice A has a different value, and normals from different circular zones of the non-spherical surface of revolution t meet at different points on the axis. They all touch a caustic surface with a cusp at the so-called paraxial image, formed by rays very close to the axis. This is the phenomenon of spherical aberration.

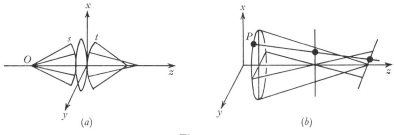

Fig. 10

In an astigmatic lens there is no longer axial symmetry, but we do assume that A is negligible. This situation is common in the human eye, and is also encountered if the object O lies off the axis of a symmetrical lens. The equation of t is

$$z = \frac{x^2}{2R_1} + \frac{y^2}{2R_2}. \tag{6}$$

This is the elliptic paraboloid (5), and we assume that R_1 exceeds R_2. We consider the elliptic zone lying in a plane at distance d from the origin. As θ varies the ellipse is described by the point P

$$\{\sqrt{(2R_1 d)} \cos \theta, \quad \sqrt{(2R_2 d)} \sin \theta, \quad d\}. \tag{7}$$

The equation of the normal to $z = f(x, y)$ at (x_1, y_1) is given by

$$(x - x_1) \Big/ \frac{\partial f}{\partial x_1} = (y - y_1) \Big/ \frac{\partial f}{\partial y_1} = (z - z_1) \Big/ -1,$$

59

and applied to (7) this gives

$$\frac{x-\sqrt{(2R_1 d)}\cos\theta}{\sqrt{(2d/R_1)}\cos\theta} = \frac{y-\sqrt{(2R_2 d)}\sin\theta}{\sqrt{(2d/R_2)}\sin\theta} = \frac{z-d}{-1}. \qquad (8)$$

If we put $z = d+R_1$ in (8) we find that $x = 0$ for all values of θ. The normals to our elliptic section of (6) thus cut a horizontal straight line that itself intersects the z axis at distance R_1 from the centre of the ellipse. Putting $z = d+R_2$ gives $y = 0$, so the normals also cut a vertical line at distance R_2. These normals form the astigmatic surface shown in Figure 11.

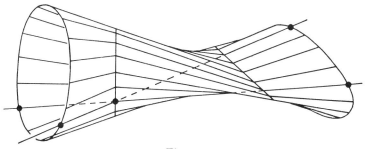

Fig. 11

This surface is of degree 4, and through each point of its 2 focal lines pass 2 generators. One such pair have been dotted in over regions where they are concealed by the surface. Optically we get 2 perpendicular line images of a point object at distances $(d+R_2)$ and $(d+R_1)$ from the lens. These lines move as d, and therefore the zone of t being considered, vary. However, the maximum values taken by d is small, so this broadening of the 2 line images is negligible.

Such a lens gives a sharp image of a vertical object at the first line and an equally sharp image of a horizontal object at the second. At the intermediate distance

$$z = d+\sqrt{(R_1 R_2)},$$

it is easily proved that normals to the zone at distance d from the origin cut a circle of radius

$$\sqrt{(2d)}(\sqrt{R_1}-\sqrt{R_2}),$$

with its centre on the axis, and in a plane perpendicular to the axis. The largest d, corresponding to rays from the circumference of the lens, defines the so-called circle of least confusion.

5. *The pitch of a ruled surface*

We first consider the lines touching a twisted curve in 3-space. They form a surface consisting of two distinct leaves intersecting in the curve itself. In Figure 12 one of the tangent lines has been emphasised; the portions on either side of the point of contact lie one in each leaf of the surface. A plane cuts the surface in a curve with a cusp lying on the original space curve, which is therefore called a cuspidal edge.

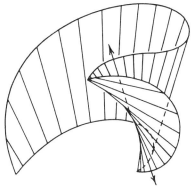

Fig. 12

Such a surface is said to be developable. Its characteristic property is that, as we move along any of its generating lines, the tangent plane remains fixed. The surface can be flattened out on to a plane without distortion.

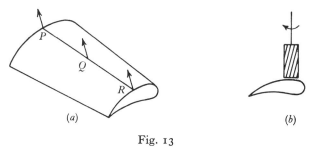

(a)   (b)

Fig. 13

In Figure 13(a) an aeroplane wing is sketched. Until recently such wings were portions of developable surfaces. Thus, besides being ruled by lines like PQR, the normals at points P, Q, R are of fixed direction.

61

This makes it easy to machine model wings for wind tunnel tests. The roughly cut shape is set up with the tangent plane at points P, Q, R horizontal. A tool rotating about a vertical axis produces a flat along this line as the shape moves perpendicular to the picture plane. After producing a series of such flats, the ridges between them are removed by hand scraping.

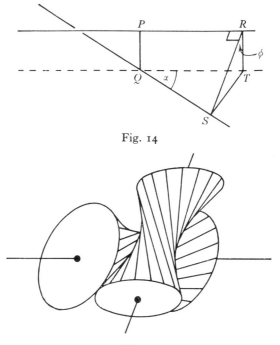

Fig. 14

Fig. 15

In the general ruled surface the tangent plane rotates as we move along a generator. In Figure 14 PR and QS are neighbouring generators, their closest distance PQ being of length d. The tangent plane at Q contains QS and PQ, while that at S contains QS and RS, R being the point of PR nearest to S. Since the angle α is small we have

$$\tan \phi \simeq \frac{\alpha \cdot QS}{d} = QS \times \text{pitch}, \quad \text{where} \quad \text{pitch} = \frac{\alpha}{d}.$$

Thus pitch relates rotation of the tangent plane to distance moved along a generator. Measured from the tangent plane at the point of striction, where neighbouring generators are closest, ϕ goes from $-\frac{1}{2}\pi$ to $\frac{1}{2}\pi$ as we move from $-\infty$ to ∞ along the generator. For a developable surface the pitch is everywhere zero, i.e. the quantity α is of second order in d.

62

The locus of points like P is called the line of striction; for the hyperboloid of Figure 3(b) it is the ellipse lying in the plane $z = 0$. If the hyperboloid is a solid of revolution, i.e. if $a = b$, it can be proved that the pitch equals c.

In Figure 15 the hyperboloids of revolution touch along a common generator and the points of striction coincide. For this to be possible the tangent planes must rotate at the same rate along the generator, i.e. the values of 'c' are equal. If the values of 'a' are also equal the surfaces are congruent, and the common generator makes equal angles with the lines of striction. Rotation about their axes then produces a pure rolling motion of one on the other.

If the values of 'a' differ pure rolling would cause points of striction to separate. Rotation about the axes now produces rolling combined with sliding of one generator along the other to maintain strictional contact. In spite of this sliding, pairs of 'hypoid' gears with teeth cut along generators are often used to couple non-intersecting shafts. The use of different values of 'a' enables a speed change to be effected as well. Hilbert and Cohn-Vossen[1] discuss these kinematic problems in greater detail.

References

1. D. Hilbert and S. Cohn-Vossen. *Geometry and the Imagination* (Chelsea, 1952).
2. D. M. Y. Sommerville. *Analytical Geometry of Three Dimensions* (Cambridge, 1934).

7

Random walks

1. *Coin tossing and random walks*

We assume that the penny being tossed has a probability p of giving heads and $(1-p)$ or q of giving tails. This implies that, in a large number of throws N, there will be about Np heads. We shall be interested in the difference between the observed number of heads h and the expected number np, where n is not necessarily large.

In Figure 1 we illustrate a one-dimensional random walk. Starting at the origin steps are of unit length to left or right; at each stage the toss of a coin determines the next step, heads indicating moves to the right.

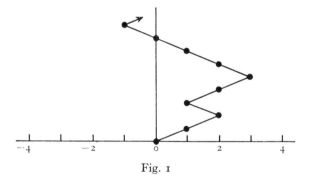

Fig. 1

Although along a line, it is easier to illustrate the walk if time is introduced as a second coordinate. Thus the walk above results from the sequence of moves

$$R\,R\,L\,R\,R\,R\,L\,L\,L\,L\,R \ldots.$$

The final distance to the right after n moves will be $(2h-n)$ if h heads have appeared.

We shall be concerned with the probability of returning to the zero line, and in the proportion of the time spent on one side. For a symmetrical penny with $p = 0.5$, we might suppose that in most walks of n steps about half the time is spent on the right. However, this is not so; in the majority of walks much more time is spent on one side than on the

64

other. In order to eliminate positions on the zero line when doing such calculations we adopt the following convention: a point on the zero line is regarded as being on the right if the previous position was on that side and vice versa.

The symmetrical two-dimensional walk involves a step to north, south, east or west at each stage, the probability of each being $\frac{1}{4}$. An analogous three-dimensional walk can also be envisaged. The probability of a return to the starting point is again of interest.

A modified two-dimensional walk inside the region bounded by a closed curve can also be considered. The walk lasts until the curve is reached, and it is therefore called an absorbing barrier. This problem is related to the solution of the Laplace equation in potential theory.

Historically the study of random walks originates from the discovery of Brownian motion. Small particles suspended in a liquid exhibit a curious zig-zag motion, even when every precaution is taken to eliminate convection currents. The explanation in terms of the bombardment of the particle by molecules of the liquid is as follows. At any instant there are a large number of impacts occurring in random directions. On average these balance out, but occasionally a preponderance of impacts in a given direction does occur. There is then a resultant acceleration in that direction.

Einstein published a theory of the Brownian motion in 1905, the year in which he announced his discovery of Special Relativity. In that year too he published a classical paper on the photoelectric effect; this, like the one on Brownian motion, introduced probability considerations into mathematical physics. Since then such considerations have become of increasing importance, and the study of random walks arises in numerous physical problems.

2. *The binomial distribution*

If a penny is thrown n times, what is the probability that just h heads appear? Such an outcome implies a sequence such as

$$H\,H\,T\,H\,T\,T\ldots H\,T,$$

containing heads h times. The probability of getting exactly this sequence is given by the product

$$p \cdot p \cdot q \cdot p \cdot q \cdot q \cdot \ \ldots \ \cdot p \cdot q,$$

where each H leads to a factor p and each T to a q. Since there are h p's and $(n-h)$ q's this probability is

$$p^h q^{n-h}.$$

However, other patterns also give h heads, in fact the total number of such patterns will be

$$\binom{n}{h} = \frac{n!}{h!(n-h)!},$$

the number of ways of choosing h out of n places for the heads to occur. As all these patterns have the same probability of occurring, the probability of h heads is

$$\binom{n}{h} p^h q^{n-h}. \tag{1}$$

The following table is for $p = 0.5$ and $n = 10$:

Table 1

No. of heads	Probability
0 or 10	0·0010
1 or 9	0·0098
2 or 8	0·0440
3 or 7	0·1172
4 or 6	0·2050
5	0·2460

From this table we can find the probability of getting less than 3 heads; it is 0·0548, obtained by adding the first 3 terms.

If we multiply (1) by t^h and sum from 0 to n, we obtain the binomial expansion of

$$(q + pt)^n, \tag{2}$$

and call (2) the probability generating function. It enables us to find the average or mean number of heads in n throws. This mean value is obtained by multiplying (1) by h and summing for h from 0 to n, giving

$$0 \cdot q^n + 1 \binom{n}{1} pq^{n-1} + 2 \binom{n}{2} p^2 q^{n-2} + \ldots + np^n. \tag{3}$$

This is a weighted sum of the numbers from 0 to n, the weights, which add up to 1, being the probabilities of occurrence of the numbers they multiply. If we differentiate (2) with regard to t and put $t = 1$ we get np. On the other hand if (2) is expanded before differentiating and putting $t = 1$ we get (3). Thus the mean number of heads is np.

66

Much the same technique enables us to find the mean value of $(h-np)^2$, two differentiations are needed and the result is npq. It is useful to note that observed values seldom differ from the expected number np by more than twice the square root of npq. In other words most trials give between

$$np - 2\sqrt{(npq)} \quad \text{and} \quad np + 2\sqrt{(npq)} \tag{4}$$

heads. Interesting too is the fact that the spread of values implied by (4) varies as $\sqrt{n}$. If we use h/n to estimate an unknown p, the uncertainty in the answer varies inversely as $\sqrt{n}$.

3. *Approximating to the binomial distribution*

In principle, (1) enables us to evaluate probabilities, or sums of probabilities, in simple coin tossing. For large n these calculations are not convenient, and we examine two well-known approximations.

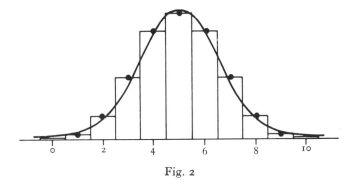

Fig. 2

In Figure 2 the values of Table 1 have been plotted as ordinates. Corresponding to the ordinate at 2, a column of the same height, and with base from 1·5 to 2·5, has been drawn and so on for all ordinates. The probability of obtaining 2 heads or less is the sum of the heights of the first 3 ordinates, or the area of the histogram taken from −0·5 to +2·5.

We see that the histogram is approximately bell-shaped, and this led de Moivre to relate it to the bell-shaped curve $\exp(-x^2)$. More precisely, he showed that, if we define a variable x by

$$x = \frac{h-np}{\sqrt{(npq)}}, \tag{5}$$

the curve

$$y = \frac{1}{\sqrt{(2\pi)}} e^{-\frac{1}{2}x^2}, \tag{6}$$

is close to the histogram. To approximate to the area under the histogram up to 2·5 we put $h = 2·5$ in (5) giving $-1·58$. Then the integral of (6) from $-\infty$ to $-1·58$ gives 0·057, a value quite close to 0·0548 found from Table 1.

The expression (6) is called the normal probability function, and for large n with p not too far from 0·5 it provides a good approximation to the binomial distribution (1). Since many applications of importance involve a large n with a small p, it is necessary to have a separate approximation for such cases. Let us consider an unusual event such as the issue of an unperforated sheet of postage stamps. The number of sheets n issued per year is large, and p, the probability that a given sheet is faulty, is small. In practice we do not usually have access to either number, but instead may well know $m = np$, the average number of faulty sheets issued per year.

We write the generating function (2) in the form

$$(q+pt)^n = (1-p+pt)^n = [1+(m/n)(t-1)]^n.$$

When n tends to infinity the limit of this expression is known to be $\exp(mt-m)$. This, when expanded, gives

$$e^{-m}e^{mt} = e^{-m}+me^{-m}t+(m^2/2!)e^{-m}t^2+\dots.$$

The coefficient of t^r in the generating function is the probability of r occurrences, just as it was in expression (2). So from the series this probability must be

$$\frac{m^r e^{-m}}{r!}. \tag{7}$$

This result is known as the Poisson distribution of probabilities.

To illustrate its use let us suppose that $m = 5$. What is the probability that, in a given year, less than 3 faulty sheets will be issued? The answer is

$$e^{-5}+\frac{5}{1!}e^{-5}+\frac{25}{2!}e^{-5} = 0·125,$$

so that for 1 year in 8 there will be 2 or fewer faulty issues.

While the Poisson distribution might be expected to apply to many rare events, it is often found that results observed do not agree well with its predictions. The reason is that it assumes a constant p in all trials. In practice a faulty perforating machine might go undetected for a period. During this time the value of p would be much higher than usual. Such fluctuations of p usually lead to a greater scatter of observed values about the mean than that predicted by the Poisson formula.

It is interesting to note an analogue of (4) for the Poisson distribution. The limits

$$m - 2\sqrt{m} \quad \text{and} \quad m + 2\sqrt{m},$$

include the majority of observed numbers of an event in a given period, where m denotes the average number in that time. Thus, if the average rate of occurrence of a particular type of industrial accident is 100 per year, we should expect fluctuations over the range 80 to 120. As indicated above, observed values may show still wider swings because the simple Poisson model does not describe the situation adequately.

4. *Returns to zero in the random walk*

We consider a random walk of $2n$ steps, and define 2 probabilities.

u_{2n} = probability of reaching the zero line at $2n$th step.

v_{2n} = probability of reaching the zero line for the first time at the $2n$th step.

The first of these involves n left and n right moves and hence

$$u_{2n} = \binom{2n}{n} p^n q^n = \binom{-\frac{1}{2}}{n} (-4pq)^n.$$

The second form for u_{2n}, obtained by algebraic manipulation, leads to

$$U(x) = 1 + u_2 x^2 + u_4 x^4 + \ldots = (1 - 4pq x^2)^{-\frac{1}{2}}, \tag{8}$$

the generating function for the u's. We define also

$$V(x) = v_2 x^2 + v_4 x^4 + \ldots .$$

The generating function $U(x)$ was constructed from known probabilities. With its aid we shall determine $V(x)$, then its series expansion will give us the values of the v's. A return to the zero line at the $2n$th step can arise in the following different ways.

(1) No previous return to zero.

(2) First return to zero at $(2n - 2)$, and a return 2 after.

(3) First return to zero at $(2n - 4)$, and a return 4 after.

....

....

(n) First return to zero at 2, and a return $(2n - 2)$ after.

Figure 3 illustrates event number (3) in this list. The dotted path includes no returns to zero. There are numerous ways in which, after the return at $(2n-4)$, we return again at $2n$. Two of these are shown, one involving also a return at $(2n-2)$.

The probability of event (3) is the product of the probability of a first return at $(2n-4)$, and of a further return in 4 steps, i.e. it is

$$v_{2n-4}u_4.$$

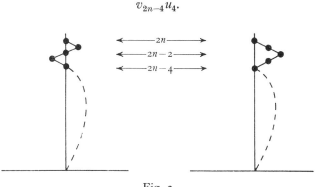

Fig. 3

We note that both the types of path shown in the figure are covered by the term u_4. We compute the probabilities of each of the events (1) to (n) and add them to get

$$u_{2n} = v_{2n} + v_{2n-2}u_2 + v_{2n-4}u_4 + \ldots + v_2 u_{2n-2}. \tag{9}$$

Consider the coefficient of x^{2n} in the product $U(x)V(x)$; for $n > 0$ this is given by the right-hand side of (9). It follows that

$$U(x) = 1 + U(x)V(x),$$

and we get at once

$$V(x) = 1 - \frac{1}{U(x)} = 1 - (1 - 4pqx^2)^{\frac{1}{2}}.$$

The binomial expansion now gives

$$v_{2n} = -\binom{\frac{1}{2}}{n}(-4pq)^n,$$

and we can evaluate the v's.

The probability of at least 1 return in $2n$ steps is

$$v_2 + v_4 + v_6 + \ldots + v_{2n},$$

since there must be a first return, and this can occur at steps 2, 4, ..., $2n$.

70

If n is large this sum is virtually equal to $V(1)$. The probability of an eventual return to zero in a long walk is thus

$$V(1) = 1 - (1 - 4pq)^{\frac{1}{2}} = 1 - [(p+q)^2 - 4pq]^{\frac{1}{2}} = 1 - |p - q|.$$

For a symmetrical walk $p = q$, and the probability of eventual return is unity, i.e. return to zero is certain.

In the two-dimensional symmetrical walk, return to the starting point is also certain. In three-dimensions however, the probability of eventual return is only 0·35. Dynkin and Uspenskii[1] give a very interesting account of two-dimensional walks, particularly for the case of a bounded region. The results just mentioned are discussed by Feller[2].

5. The time spent on one side

Feller[2] proves the results of this section; they are discussed here because of their rather unexpected nature. We consider a walk of $2n$ steps, and, with the allocation of points on the zero line adopted earlier, there will be $2r$ positions on the right and $(2n - 2r)$ on the left. The probability of this event can be shown to be

$$\binom{2r}{r}\binom{2n-2r}{n-r}\frac{1}{2^{2n}}. \tag{10}$$

The method of proof is very similar to that used in the last section, and the formula for v_{2n} found there is a central requirement.

The expression (10) gives the values of Table 2 if $n = 10$ (20 steps).

Table 2

$2r$	Probability
0 or 20	0·1762
2 or 18	0·0927
4 or 16	0·0736
6 or 14	0·0655
8 or 12	0·0617
10	0·0606

Thus the least likely outcome is that half the time be spent on each side. It is nearly 3 times as likely that all the time will be on the right. Denoting the proportion of time on the right by $x = r/n$, we can prepare the frequency diagram of Figure 4. It is U-shaped, in sharp contrast to the normal curve.

The probability of being on the right for 4 or less steps, obtained by adding the first 3 terms of Table 2 is 0·3425. We seek an analytic function which, like the normal curve of Section 3, makes such computations easier.

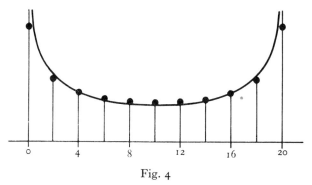

Fig. 4

Stirling's asymptotic formula is

$$n! \sim n^n \sqrt{(2\pi n)}\, e^{-n},$$

and it can be used to replace each of the factorials in (10). Simple manipulation gives

$$\frac{1}{n\pi\sqrt{[x(1-x)]}},$$

whose graph is a U-shaped curve, infinite at $x = 0$ or 1. Writing $dx = 1/n$, we find that the chance of spending a fraction X or less of the time on the right is about

$$\int_0^X \frac{dx}{\pi\sqrt{[x(1-x)]}} = \frac{2}{\pi}\sin^{-1}\sqrt{X},$$

the Arc Sine Law of Chung and Feller.

We set $X = \frac{5}{20}$ for the probability of spending 0, 2 or 4 steps on the right, a continuity correction of half an interval of size 2 being added to 4 to give the top line of X. This correction should be compared with integration up to 2·5 in Section 3, when a probability of 2 or less was being estimated. The result is 0·3333, reasonably close to our exact value of 0·3425. What is the probability of spending less than 1% of the time on the right in a long walk? The formula gives 0·064, so this apparently unlikely situation will arise oftener than once in 20 walks.

72

6. *A Monte Carlo solution of the Laplace problem*

We wish to find a function $\psi(x, y)$ at points inside a closed curve, with its value specified at each point of the boundary, and with

$$\nabla^2\psi \equiv \frac{\partial^2\psi}{\partial x^2} + \frac{\partial^2\psi}{\partial y^2} = 0, \tag{11}$$

satisfied at all internal points.

Figure 5 (*a*) shows a square mesh of points superimposed on the curve, points on the stepped boundary of this mesh lie close to the curve. The continuous problem is replaced by that of finding $V(P)$ at each internal mesh point like P. At boundary points V takes the value that ψ has at the nearest point of the true curved boundary. It remains to find a relationship between neighbouring V's analogous to (11). Let the mesh interval be h, and suppose that K, L, M and N are the nearest points to a typical mesh point J.

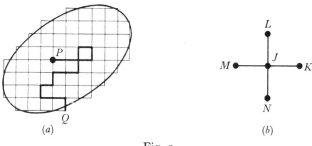

(*a*)   (*b*)

Fig. 5

We approximate as follows

$$\left(\frac{\partial V}{\partial x}\right)_{u-\frac{1}{2}h} \simeq \frac{V(J) - V(M)}{h}, \quad \left(\frac{\partial V}{\partial x}\right)_{u+\frac{1}{2}h} \simeq \frac{V(K) - V(J)}{h},$$

where (u, v) are the coordinates of J. These in turn give

$$h^2\left(\frac{\partial^2 V}{\partial x^2}\right)_u \simeq h\left(\frac{\partial V}{\partial x}\right)_{u+\frac{1}{2}h} - h\left(\frac{\partial V}{\partial x}\right)_{u-\frac{1}{2}h} \simeq V(M) + V(K) - 2V(J).$$

A similar result in the y direction leads us to replace (11) by

$$4V(J) = V(K) + V(L) + V(M) + V(N). \tag{12}$$

There is a value of V to be found at each internal point, and each such point also supplies an equation like (12). While a direct solution of the set of equations is feasible, a number of iterative ways of approximating to the answer have been developed.

We adopt another approach; to find $V(P)$ we conduct a large number of symmetrical walks from P. These terminate as soon as a boundary point Q is encountered. If n_i walks out of a total of N terminate at the ith boundary point, where V takes the known value V_i, then

$$V(P) \simeq \sum_{(i)} \frac{n_i V_i}{N},$$

the sum being over all boundary points.

The justification is as follows. We define $U_i(P)$ as the probability that a walk starting at P ends on the ith boundary point. Then the function U_i takes the value 1 at the ith boundary point and zero at all others. Moreover, the probability of moving from J to each of K, L, M and N is $\frac{1}{4}$, for our walk is symmetric. Thus the probability of ending at the ith boundary point after starting from J is given by

$$U_i(J) = \tfrac{1}{4}U_i(K) + \tfrac{1}{4}U_i(L) + \tfrac{1}{4}U_i(M) + \tfrac{1}{4}U_i(N).$$

Comparison with (12) shows that the function U_i is also an approximate solution of the Laplace problem with the boundary values defined above. Our random walks enable us to estimate $U_i(P)$ by

$$U_i(P) \simeq \frac{n_i}{N}.$$

We next consider the function whose value at P is defined by

$$W(P) = V_1 U_1(P) + V_2 U_2(P) + \ldots + V_i U_i(P) + \ldots, \tag{13}$$

the sum being over all boundary points. Its value at the ith boundary point is V_i, for there all U's except U_i vanish, and this one is unity. So $W(P)$ has the same boundary values as $V(P)$. Moreover, as

$$\nabla^2 W(P) = \sum_{(i)} V_i \nabla^2 U_i(P) \simeq 0,$$

this function satisfies the Laplace equation in finite difference form. It is known that the problem has a unique solution, so that $W(P)$ coincides with $V(P)$. We estimate it by

$$V(P) = \sum_{(i)} V_i U_i(P) \simeq \sum_{(i)} \frac{n_i V_i}{N}.$$

While such a method would certainly not be used in this case, it does illustrate an approach of great value. In complicated physical situations

Monte Carlo procedures of this kind are often the only possible way of obtaining solutions. With the advent of the electronic computer they have been widely used in such fields as nuclear physics.

References

1. E. B. Dynkin and V. A. Uspenskii. *Random Walks* (D. C. Heath and Co., 1963).
2. W. Feller. *An Introduction to Probability Theory and its Applications*, vol. 1, 2nd edition (Wiley, 1957).

8

The four-colour problem

1. *Introduction*

Cayley proposed the problem of determining the minimum number of colours needed for a map, no two adjacent regions being of the same colour. The sea surrounding the land area is also to be coloured; Figure 1 (*a*) is of a simple map needing 4 colours. The colours marked 2 do not count as touching at the multiple vertex of Figure 1 (*b*). Such vertices can always be removed as in (*c*), and the resulting map, with only triple vertices, is called regular. A solution for the transformed regular map evidently gives a solution for the original one, but the converse does not hold. All areas are assumed to be simply connected, i.e. there are no ring-shaped regions.

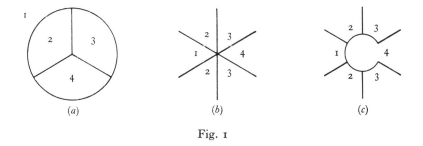

(*a*)   (*b*)   (*c*)

Fig. 1

In spite of much effort over the last century the minimum number of colours is still not known. We saw above that 4 are necessary, and we prove later that 5 suffice. We shall also show that, if there are less than 12 regions, 4 colours will do. Various workers have gradually increased this number from 12 to 39. A number of equivalent statements of the problem can be formulated; we shall discuss one later. The 2- and 3-colour problems are easier and we shall derive the conditions under which these smaller numbers are sufficient.

We can think of our map as lying on a sphere; in fact the problem is a topological one. This means that any continuous distortion of the

76

spherical surface without tearing does not affect its solution. However, the problem of a map lying on an anchor ring is different. Oddly enough it is easily solved; we shall prove that 7 colours always suffice. It is possible to draw 7 hexagons on the anchor ring, each of which touches the other 6. Thus 7 colours are necessary for this map; we next examine its derivation.

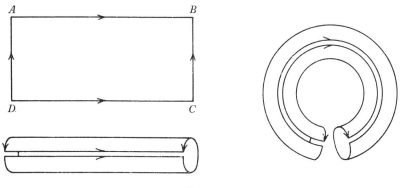

Fig. 2

Figure 2 shows how an anchor ring can be formed from a rectangular sheet of rubber. This is first rolled into a cylinder so that edges AB, CD come into contact. Then the cylinder is bent into a ring and the circular

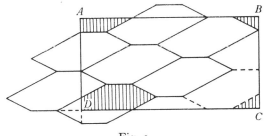

Fig. 3

ends joined. We note that the arrows on these end circles run in the same direction. The possibility of a closing up of the rubber sheet with a reversal of directions is discussed by Hilbert and Cohn-Vossen[4].

In Figure 3 we have a hexagon surrounded by 6 others. Various parts of the cluster lie outside the rectangle $ABCD$, but they have counterparts within this area indicated by dotted lines. As a consequence, folding of $ABCD$ as in Figure 2 transfers the cluster of 7 hexagons to an anchor ring. We note, for instance, that the region containing D is built up of

77

four separate shaded portions, one at each vertex of the rectangle. We see too that, because of the way it is formed on the ring, it touches the other 6 hexagons; and this is true for all 7 hexagons.

We consider next a solid figure-of-eight, or alternatively a sphere with two non-intersecting holes bored through it. This problem, like that of the sphere with one hole, i.e. the anchor ring, is also solved. Eight colours always suffice, and some maps cannot be coloured in less. Similar results have been found for many other surfaces obtained by this hole-boring process.

2. Euler's theorem

We shall need Euler's formula

$$F + V - E = 2, \qquad (1)$$

for a map of F regions with V vertices and E edges. This was given as a result for convex polyhedra in Chapter 1. Figure 4 illustrates how a cube can be opened out into a plane map. The face $EFGH$ is removed and the 4 faces attached to $ABCD$ stretched till they fold out flat. The

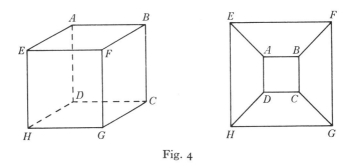

Fig. 4

resulting map has the same edge and vertex numbers as the cube. The missing face $EFGH$ can be taken as corresponding to the outside of the map so that both map and cube have the same F number. A mapping of this kind is possible for any convex polyhedron, so that result (1) applies equally to the 2 structures. We shall prove it for plane maps, and start by adding lines to triangulate all regions with more than 3 sides. For an n-sided region we can join any one vertex to $(n-3)$ others, thus creating $(n-3)$ new edges and $(n-3)$ new regions. As a result the left-hand side of (1) has not been altered.

78

We now work round the outer edge of the map removing one triangle at a time. Figure 5 illustrates the three possible ways in which a triangle is eliminated. In Figure 5(a) we remove edge AB losing also one region, so that $(F+V-E)$ is unaltered. In Figure 5(b) we delete edges AB and BC, also losing a region and a vertex, so again $(F+V-E)$ is unchanged. In Figure 5(c) we lose 3 sides, 2 vertices and 1 region, so that once more the left-hand side of (1) is unchanged. We carry out removals until a single triangle remains, then $E = 3$, $V = 3$, $F = 2$ (the outside area being counted as a region). So the left-hand side of (1) must be 2, and our proof is complete.

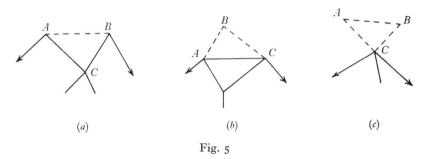

(a)   (b)   (c)

Fig. 5

For a map on a sphere with holes (1) is replaced by

$$F+V-E = K, \tag{2}$$

and K is called the characteristic of the surface. It is 2 for the sphere, o for the anchor ring, -2 for the figure-of-eight and $(2-2n)$ for a sphere with n holes. It is of interest to note that there is a series of one-sided surfaces for which K takes odd values; these are discussed by Arnold[1] and Hilbert and Cohn-Vossen[4]. An ingenious proof of (2) for the anchor ring is given by Goodstein.[3]

In the special case of a regular map we have 3 edges at each vertex. A count of edges gives

$$3V = 2E, \tag{3}$$

the 2 allowing for the fact that each edge is counted twice in $3V$. Suppose there are n_3 triangles, n_4 quadrilaterals, and so on, then

$$F = n_3+n_4+n_5+.... \tag{4}$$

Counting edges by face-polygons gives

$$2E = 3n_3+4n_4+5n_5+.... \tag{5}$$

From (2), (3), (4) and (5) we deduce the useful result

$$- 3n_3 - 2n_4 - n_5 + n_7 + 2n_8 + \ldots \ = \ -6K, \qquad (6)$$

valid for regular maps.

3. The five-colour theorem

Results in this section will be proved for regular maps; as remarked earlier they then hold for any map. Our first argument shows that 6 colours suffice for the plane map. The same method applied to the anchor ring gives 7 colours. We then apply a subtler argument to show that, for the plane, 5 colours will always be adequate.

From equation (6) with $K = 2$ (the plane) we see that at least one of the numbers n_3, n_4 or n_5 is non-zero, as otherwise the left-hand side would not be negative. Suppose that n_5 is not zero, so there is at least one pentagon P. Remove the dotted side, as in Figure 6(a). The resultant map has one region less and we assume that 6 colours suffice for it.

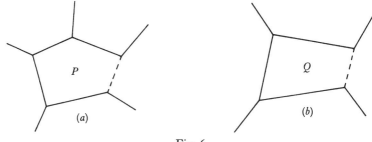

Fig. 6

At most, 4 of these will touch P, with one of them inside it. We can restore the dotted side, leave all colours outside P unchanged, and paint P with the remaining colour. This supplies the basis of an inductive proof, for we have reduced the problem of colouring a map of n regions in 6 colours to the same problem for a map of $(n-1)$ regions.

Had there been no pentagons, but at least one quadrilateral Q, the second diagram indicates how the same procedure could be used. Similarly, if there are no P's or Q's, a triangle must be present, and can be eliminated.

If $K = 0$ (the anchor ring) the left-hand side of (6) is zero and two cases arise. When a region of 7 or more sides occurs there is at least one of 5 or fewer, and vice versa. Otherwise the map must consist only of hexagons. A hexagon can be eliminated in this case, and we assume the remaining map can be coloured in 7 colours. Only 6 touch the eliminated hexagon, the side can be restored and the seventh colour inserted in the hexagon. Again, an inductive proof follows, and if there are no hexagons, there must be regions of 5 or fewer sides. These can be eliminated in the same way, so that, at any stage, the problem can be reduced to one for a map with one region less.

Returning to plane maps we can show that, in Figure 7(a), at least one of the pairs of regions A, D and B, C are not in contact. For if A touches D then region B is surrounded by the chain A–P–D and cannot touch C or vice versa. This result does not hold for the anchor ring.

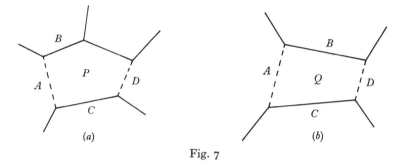

Fig. 7

Remove the boundaries separating A and D (assumed not to be in contact) from P. Assume that 5 colours suffice for the new map; only 4 will be involved around P, and one of these will belong to A, P and D. We restore the boundaries and leave all colours unaltered outside P. Since A and D, now separate regions again, do not touch, this restoration does not upset the basic colouring condition. We can use the fifth colour for region P. The second diagram applies when there are no pentagons, but at least one quadrilateral; a triangular region is removed by deleting one side. Again an inductive proof, also for 5 colours, follows.

Finally, consider a plane map with less than 12 regions. From (6) there must be at least one quadrilateral or one triangle, since n_5 is less than 12. Taking the case of a Q, as in the second diagram of Figure 7, then one of the pairs A, D or B, C are not in contact. Deleting sides on the assumption that A, D do not touch, the reduced map is assumed to be painted in 4 colours, of which only 3 touch Q. The basis of the induction follows as before.

4. Heawood's formula

Heawood found a remarkable formula for all surfaces with $K \leqslant 0$. He showed that

$$\tfrac{1}{2}[7 + \sqrt{(49 - 24K)}], \tag{7}$$

colours suffice for such surfaces, the square brackets denoting that the integer part of the expression inside is to be taken. If $K = 0$ the formula gives 7, and we have proved that some maps require this number. In many other cases it has been shown that (7) gives the best possible number.

To derive (7) we start with (6) written in the form

$$\sum_{j=3}^{\infty} (j - 6)n_j = -6K, \tag{8}$$

and assume the existence of a map whose r-coloration cannot be reduced to that of a map with one region less in r colours. We deduce an upper bound for r. If this upper bound is exceeded, our assumption is false, and all maps can be reduced. Thus an inductive colouring is possible.

If the map has a region of $(r-1)$ or fewer sides, the method of the previous section shows that its colouring can be reduced to that of a map with one region less. Hence for all j with non-vanishing n_j

$$j > r - 1 \quad \text{or} \quad j - 6 \geqslant r - 6. \tag{9}$$

Moreover, the map must have more than r regions or there is an immediate colouring in r colours, so

$$\Sigma n_j \geqslant r + 1. \tag{10}$$

Expressions (8), (9) and (10) lead to

$$-6K = \Sigma(j - 6)n_j \geqslant (r - 6)\,\Sigma n_j \geqslant (r - 6)(r + 1),$$

the second result depending on the assumption that r exceeds 5. This in turn implies that $K \leqslant 0$; the inequality gives

$$r^2 - 5r + 6(K - 1) \leqslant 0. \tag{11}$$

Provided that $K \leqslant 0$ (11) is satisfied for $r = 0$, and remains valid until

$$r > \tfrac{1}{2}\{5 + \sqrt{(49 - 24K)}\}. \tag{12}$$

Adding 1 to this quantity, and taking the integer part we get (7). This number of colours causes (11) to fail, and by the above reasoning an inductive r-coloration is possible. We made two assumptions in deriving this result; that $r > 5$ and that $K \leqslant 0$. The smallest value taken by (7) is 7, provided the second condition holds. Thus the first assumption is not a restriction. The second is vital, as it excludes the all-important case of plane maps.

5. *The two-colour theorem*

In the case of two colours the necessary and sufficient condition on the map is that, at each vertex, there are an even number of edges.

Figure 8(a) shows a vertex with colours 1 and 2 alternating round it. This alternation is only possible for even vertices, so the condition is necessary.

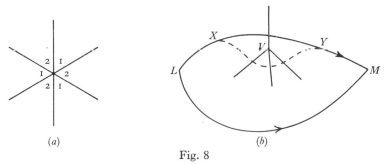

(a)   (b)

Fig. 8

Allotting colour 1 to any region we can reach any other by many paths. Selecting any one path we change colour each time we cross an edge, and eventually all regions will be coloured. For the process to succeed we must be sure that, starting from L in one region, any two paths leading to M in another, give this latter region the same colour.

Figure 8(b) shows two such paths. We can deform the upper continuously into the lower one. In the deformation we need only examine what happens each time a vertex like V is crossed. In the diagram points X and Y are separated by one edge before V is crossed and three after. They are thus of opposite colours in either case. The argument is general; if k edges are crossed before a vertex is encountered and l edges after, then $k+l = 2n$. Hence, k and l are both odd or both even. In the first case there is a colour change on both new and old routes, in the second no change occurs on either.

Thus our colouring process determines all regions uniquely once a single one has been allotted. The alternating process guarantees that adjacent regions always differ, and the problem is solved.

6. *The three-colour theorem*

We restrict ourselves to regular maps, and prove that the necessary and sufficient condition for 3 colours to suffice is that all regions have an even number of sides. The proof depends upon constructing a dual map. We select a point in each region to form the vertices of this new map. Edges of the new map cross edges of the old, and the triple vertices of the old map imply that all regions of the dual map are triangles.

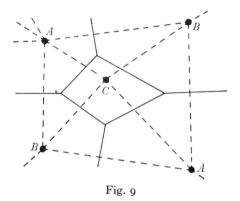

Fig. 9

A colouring of regions of the first map in 3 colours leads to a colouring of vertices of the dual map. This has the property that, for each of its triangles, the vertices carry different colours. In Figure 9 the vertex inside the quadrilateral is coloured C, and the 4 surrounding vertices alternate from A to B. We deduce a face colouring of the dual map in 2 colours from its vertex colouring by A, B and C. If A, B, C occur clockwise round a triangle it is white, otherwise black. The rotational rule means that 2 black or 2 white regions never touch.

Thus, starting with a 3-colour map, we derive a 2-colour solution for the dual map. For this to be possible all its vertices must be even by Section 5. Hence it is necessary that all regions of the 3-colour map have an even number of sides.

To prove the converse, we note that, if all regions of the original map have an even number of sides, then the dual map, having only even vertices, can be coloured in 2 colours. We allot a clockwise direction to white and anticlockwise to black. Starting with a white triangle we place colours A, B and C clockwise at its vertices. The 3 adjacent triangles are

84

black and the remaining vertex of each is coloured to produce anti-clockwise ordering. This process is continued until all the vertices are coloured. There can be no contradictions because of the fact that adjacent triangles always carry opposite colours, and therefore opposite directions of rotation. Finally, we observe that the vertex colours of the dual map can be regarded as face colours of the original. At any of its triple vertices each of the colours A, B and C occur, so adjacent colours always differ.

7. The Tait–Wolinskii theorem

Tait showed that, for regular maps, the 4-colour problem is equivalent to an edge-colouring problem using 3 colours, so that, at each vertex, the 3 edges differ. Wolinskii, a young Russian killed in World War 2, rediscovered this result, and gave the following elegant proof.

We first show that a 4-colour solution for faces leads to a 3-colour solution for edges. Let the face colours be denoted by the 4 number pairs

$$(0, 0), \quad (0, 1), \quad (1, 0), \quad (1, 1). \tag{13}$$

We define an addition operation for pairs of these symbols using addition modulo 2, for instance

$$(0, 1) + (1, 1) = [1, 2] = [1, 0].$$

We readily verify that the 6 pairs give 3 different sums

$$A = [0, 1], \quad B = [1, 0], \quad C = [1, 1]. \tag{14}$$

Edges are coloured A, B or C, the colour being determined by adding face symbols on either side of an edge.

If we go round any vertex adding the edge symbols and using the same rule adopted for face symbols we obtain the number pair $(0, 0)$. For, by the definition of the edge symbols, this is equivalent to adding together the 3 face symbols twice each, and this must give $(0, 0)$ because of modulo 2 addition, whichever faces colours are involved. Now, if 2 of the edge symbols at this vertex coincide, their sum is $(0, 0)$, again because of modulo 2 addition. Thus the sum of all 3 could not be $(0, 0)$, and coincidence of edge symbols at a vertex is ruled out, i.e. we have obtained a proper edge colouring.

We now have to deduce a face colouring in 4 colours from an edge colouring in 3. Starting at any face we allot colour $(0, 0)$ and take any route we like from face to face. On leaving face coloured (p, q) we cross edge coloured $[E, F]$. The colour of the new face entered is given by

$$(r, s) = (p, q) + [E, F] = (p+E, q+F),$$

with our usual addition rule.

We have to ensure that this process allots colours uniquely, and consider the path deformation shown in Figure 10. The upper path from L to M on its way into the lower will cross a number of vertices like V. On the old path the addition to face colour symbol on going from X to Y is one of the pairs A, B or C. On the new path it is the sum of the other 2. The sum of the number pairs A, B and C is (0, 0), and hence, remembering our modulo 2 addition, the sum of any 2 equals the third. Thus the colour change on going from X to Y is the same by either path, and our proof is complete.

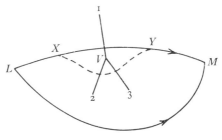

Fig. 10

An interesting consequence of this theorem is that 4 colours suffice for any regular map all of whose regions have their numbers of sides divisible by 3. We can colour the edges of such a map in 3 colours as follows. Allot any colour to the first edge and then follow various paths along edges. If the colours are denoted by 0, 1, 2, then when we encounter a new edge on the path being pursued its colour is decided as follows. At its junction with the previously coloured edge a third edge will lie either to right or left of the path. If to the right add 1 to the old colour number, if the left subtract 1. The resulting number denotes the colour for the new edge. It must be interpreted modulo 3, thus if it is −1 we add 3 to get colour 2, if it is 3 we subtract 3 to give colour 0. It can be proved (see Dynkin and Uspenskii[2]) that any 2 paths give a unique colour to a given edge, so that no contradictions will arise. As there is a 3-colour solution for edges, there is a 4-colour solution for faces.

We close by mentioning a remarkable colour theorem proved by

Dynkin and Uspenskii[2]. If a sphere has on it a 3-colour map, then there is at least one pair of points at the opposite ends of a diameter lying in regions of the same colour. Another excellent account of colouring problems is given by Stein.[5]

References

1. B. H. Arnold. *Intuitive Concepts in Elementary Topology* (Prentice-Hall, 1962).
2. E. B. Dynkin and V. A. Uspenskii. *Multicolor Problems* (D. C. Heath and Co., 1963).
3. R. L. Goodstein. *Fundamental Concepts of Mathematics* (Pergamon, 1962).
4. D. Hilbert and S. Cohn-Vossen. *Geometry and the Imagination* (Chelsea, 1952).
5. S. K. Stein. *Mathematics the Man-made Universe* (W. H. Freeman and Co., 1963).

9

Dissection problems in two and three dimensions

1. *Introduction*

The word dissection is applied to a wide range of mathematical procedures and problems. The classical dissection puzzle is typified by Figure 1, where a square has been cut into five pieces and reassembled to form a regular octagon. In such problems the emphasis is on the smallest number of pieces needed for a solution, and Lindgren[6] discusses them in detail. Later we give a proof of Bolyai's theorem, that any plane polygon can be dissected into any other of equal area.

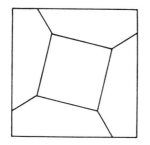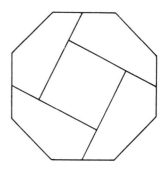

Fig. 1

Figure 2 illustrates another type of problem, that of dissecting a rectangle into unequal squares; the 33 × 32 rectangle provides 9 squares with the sides indicated. Similarly, one could seek a dissection of a square into unequal squares; Gardner[4], Stein[8] and Meschkowski[11] give excellent accounts of these tiling problems.

Another application is shown in Figure 3, based on the dissection of a square into either 4 right isosceles triangles or into 4 of these and a regular octagon. The second dissection is applied to all members of a tessellation of squares, and the triangular pieces reassembled to form the smaller squares. Other tessellations of the plane can be generated in the same way; an interesting account is given by Kraitchik[5]. We shall apply this process to solid tessellations.

In three dimensions there is no simple counterpart of Bolyai's theorem, and Dehn proved that it is not possible to dissect a regular tetrahedron into a cube of the same volume. A complete theory has appeared quite recently, clearly outlined by Boltyanskii[1]. In spite of the greater difficulty of the general theory, many ingenious puzzles in 3-space have been developed. The Soma puzzle given by Gardner[4], and the similar but more difficult cube puzzle discussed by Steinhaus[9], are worth trying. Lindgren[6] gives a 7-piece dissection of a $2 \times 1 \times 1$ block into a cube.

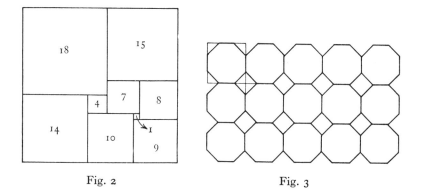

Fig. 2   Fig. 3

Figure 4 shows 2 attractive dissections of the cube into 4 congruent parts. Ehrenfeucht[3] discusses the first, for which one of the 4 parts is drawn below; we return to the second later. Here each part consists of a triangular dipyramid; for the one drawn, ABC is the common base, and P, Q the 2 vertices.

The arithmetic results

$$3^3 + 4^3 + 5^3 = 6^3,$$

$$1^3 + 6^3 + 8^3 = 9^3,$$

$$1^3 + 12^3 = 9^3 + 10^3,$$

can be illustrated by dissection. Lindgren[6] describes an 8-piece dissection of a $6 \times 6 \times 6$ cube that can be reassembled to form 3 cubes of sides 3, 4 and 5. The other two are to be found in Cadwell[2].

We conclude this introduction with a dissection proof of the fact that the sum of the cubes of the first n integers is the square of their sum. Cubes of sides 1, 2, 3, ..., n are first cut into square tiles of unit thickness.

89

From one of the 2×2 tiles a 1×1 square is removed, one of the 4×4 tiles has a 2×2 square removed and so on. Figure 5 (for $n = 4$) shows how the pieces are arranged to produce a square of unit thickness, and with side $(1 + 2 + \ldots + n)$. The use of dissection methods in proof will be illustrated further by a celebrated result due to Besicovitch.

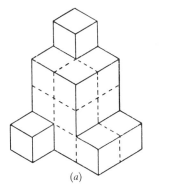 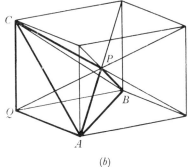

(a) (b)

Fig. 4

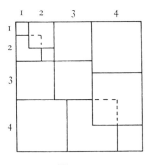

Fig. 5

2. *The two-dimensional dissection theorem*

This theorem states that the necessary and sufficient condition for dissecting polygon P into polygon Q is the equality of their areas. Hadwiger and Glur recently showed that a solution can be found so that the edges of each piece in P are parallel to its edges in Q. This implies that moving the piece from P to Q requires a parallel translation alone, or such a move together with a half-turn.

We observe from Figure 6(a) that a triangle can be dissected into three pieces that form a rectangle; both parallel translations and half-

90

turns arise in the reassembly. The second diagram shows that parallelograms on the same base and between the same parallels can be dissected into each other; no half-turns are involved here.

In Figure 7 the rectangles with diagonals AC, EG are of the same area. This will also be true of rectangles BE, CF; and we show that these can be dissected into each other From the area equalities we see that

$$\frac{DE}{DC} = \frac{DA}{DG} \quad \text{and} \quad \frac{BH}{HF} = \frac{HC}{HE}.$$

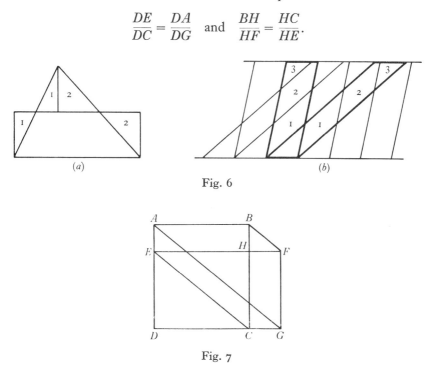

Fig. 6

Fig. 7

As a result AG is parallel to both EC and BF. Using the parallelogram dissection above we transform

> rectangle BE to parallelogram AF,
> parallelogram AF to parallelogram BG,
> parallelogram BG to rectangle CF.

We now have dissected rectangle AC into rectangle EG using only parallel translations.

We now proceed to the main theorem, and start by dividing P into triangles. Each triangle is then dissected and reassembled to form part of a column of unit width, and with height equal to the area of P. The construction for a typical triangle (No. 3) is shown in Figure 8. It is first transformed to a rectangle as above, then the rectangle is transformed to a parellelogram with its base parallel to that of the column being constructed. In the second illustration the parallelogram goes into a rectangle, and the rectangle into a new one of base unity. This rectangle is then moved into position in the column, triangle No. 4 is dealt with, and so on.

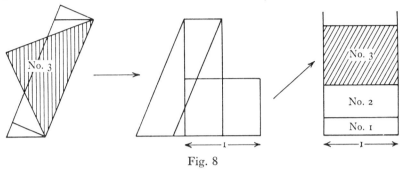

Fig. 8

A similar process applied to Q gives a column of the same dimensions. This is superimposed on the P column and all its cuts added to those already existing in P. The more numerous and smaller pieces that result can now be reassembled into either of the original polygons. Moreover, in moving these pieces only parallel translations and half-turns arise. Hence all cuts in P are parallel to their positions in Q.

3. Three-dimensional tessellations

One of our earliest geometrical experiences is the filling of space with equal cubes. Many other packings exist, and we restrict the field by allowing only one or more kinds of Archimedean solid. In addition we require each edge to be surrounded by the same set of solids throughout the tessellation. Thus for our packing of cubes each edge is surrounded by 4 cubes. Andreini proved that there are just 4 more tessellations of this type. We do not give this proof, but show how they may be derived from the cubic filling by dissection and reassembly. The 5 Andreini tessellations consist of:

(1) cubes,
(2) octahedra and cuboctahedra,
(3) truncated octahedra,
(4) octahedra and tetrahedra,
(5) tetrahedra and truncated tetrahedra.

In Figure 9(a) the dotted equilateral triangle indicates the start of the process that turns a cube into the cuboctahedron shown in Figure 11(a) of Chapter 1. This dotted section cuts off a tetrahedron, and 8 of them will form a regular octahedron. By applying the same dissection to all the cubes in a cubic space packing we can therefore derive a packing involving cuboctahedra and octahedra. The second diagram of Figure 9 shows 4 cuboctahedra stacked together. It is easy to visualise the octahedron, half of which fills the central gap in this layer.

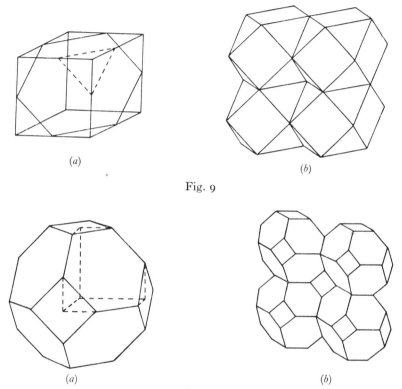

(a)

(b)

Fig. 9

(a)

(b)

Fig. 10

Returning to Figure 9(a) we see that a plane parallel to the dotted triangle can be chosen to cut the cube into two equal pieces, each with a regular hexagonal face. This section is obtained by joining the mid-points of sides of the cube's Petrie polygon, mentioned in Chapter 1.

The truncated octahedron of Figure 10(a) has 6 square faces and 8 regular hexagonal faces. The squares arise by cutting off the 6 corners of a regular octahedron in such a way that the 8 faces, originally equilateral triangles, become regular hexagons. The dotted lines indicate that this solid consists of 8 identical shapes each of the form obtained by bisecting a cube as above. This dissection of all cubes in a packing can be carried out and the resulting pieces reassembled to form truncated octahedra. A group of 4 such solids in the second diagram shows clearly that another one will fit into the central gap. A number of biological applications of this packing are discussed by D'Arcy Thompson.[10]

4. The remaining Andreini tessellations

We first note that appropriate parallel shifts will distort a cubic pack into one of parallelopipeds with rhombic faces. These shifts can be chosen to make the acute angles of all faces equal to $\frac{1}{3}\pi$. Figure 11(a) shows such a parallelopiped divided into a regular octahedron and two regular tetrahedra. This dissection leads to a method of filling space with these regular solids.

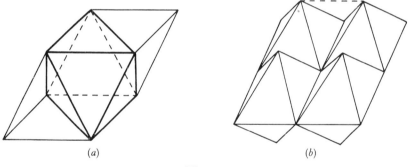

(a)   (b)

Fig. 11

The second diagram shows 4 octahedra in edge contact. The dotted line indicates how one of the gaps between them can be filled by a regular tetrahedron. If all 4 gaps on the upper surface are filled in this way a central pit remains; this will accommodate just one-half of a regular octahedron. A series of these new octahedra form the next layer. Rouse Ball[7] gives a fine illustration of this packing. We note that an octahedron of this packing, together with the 8 tetrahedra touching its faces, constitute the stella octangula shown in Figure 6 of Chapter 1. Figure 12 indicates a regular tetrahedron cut from a cube. The 4 remaining portions are each one-eighth of a regular octahedron. So this

94

mixed tessellation can also be derived from a cubic pattern without first distorting to rhombic parallelopiped form.

Figure 13(a) shows a rhombic parallelopiped cut into 2 regular tetrahedra and 2 truncated tetrahedra. The latter solid is obtained from a tetrahedron by cutting off corners suitably, it has 4 triangular and 4 hexagonal faces, all regular. The associated space filling is illustrated by a block of 5 truncated tetrahedra in contact. One of them is surrounded by the rest and is just visible in the diagram. The gaps between them are filled by the regular tetrahedra, giving another mixed tessellation.

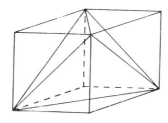

Fig. 12

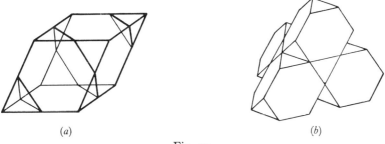

(a)   (b)

Fig. 13

5. *The rhombic dodecahedral pack*

The rhombic dodecahedron, illustrated in Figure 11(b) of Chapter 1, has 12 rhombic faces, and is the dual of the cuboctahedron. While not an Archimedean solid it provides another important tessellation of space. This is related to close packing of spheres and to the bee's cell, topics discussed by Rouse Ball[7] and D'Arcy Thompson.[10]

A cube is divided into 6 congruent square pyramids by joining its centre to each vertex. If these are stuck on the 6 faces of another cube, as in Figure 14(a), a rhombic dodecahedron results. Thus, if alternate

95

cubes in a pack are so dissected, and the bits reassembled with uncut cubes, a rhombic dodecahedral pack results. A group of 4 dodecahedra in contact is shown in the diagram.

This packing can be generated in another way, depending on the dissection of the cube shown in Figure 4(b). Eight of the resulting pieces fit together to form a rhombic dodecahedron, so another splitting of the basic cubic tessellation, this time involving all cubes, produces the same result.

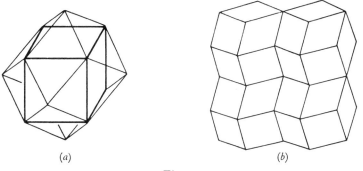

(a)   (b)

Fig. 14

6. *Rotating a rod in minimum area*

In moving a rod through angle π about its centre it sweeps out a circular area. Can the same reversal of direction be achieved, but with a smaller area? Figure 15(a) represents a three-cusped hypocycloid. It has the property of cutting off a constant length PQ along its tangents. Starting with the point of contact P at cusp 2, and rotating the tangent anticlockwise, it moves until Q touches the curve at cusp 3, with P lying at the mid-point of arc 1–2. Then P moves on to touch the curve at 1, Q being at the mid-point of arc 2–3. Finally, Q moves up to 2 and the tangent PQ has reversed its direction, sweeping out the area of the curve. For some time it was thought that this was the minimum area solution, but Besicovitch proved the surprising result that, by a suitable combination of motions, the area swept out could be made as small as required.

We first show that a rod can be moved to a parallel position sweeping out a small area. In Figure 15(b) a rotation through angle ϕ carries P_1 to P_2. The rod then moves along its length until its centre lies on the line of the new position required. This brings P_2 to P_3, and a reverse rotation of ϕ takes P_3 to P_4. Finally, a shift along the rod's length takes P_4 to P_5. The total area swept out is $2l^2\phi$, where $2l$ is the length of the rod. This can be made small if ϕ is small, but at the cost of moving a considerable distance from P_2 to P_3.

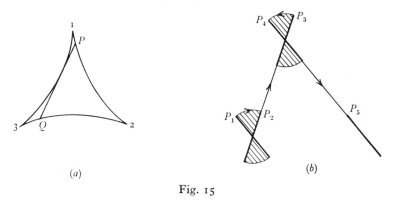

(a)

(b)

Fig. 15

Next we prove that a triangle can be dissected, and the pieces reassembled by parallel translation so as to overlap. The new total area is $2/(m+2)$ times the area of the triangle, where m is an arbitrary integer.

We illustrate the case $m = 3$, and Figure 16 shows triangle ABC divided into 2^m smaller triangles. In addition we insert $(m+1)$ guide lines at equal intervals, and parallel to its base.

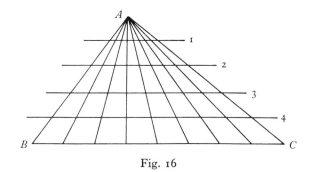

Fig. 16

The first step reassembles the 4 pairs of adjacent triangles as in Figure 17. The pairs overlap along the second guide line. We consider the total lengths they cut off on each of the 4 guide lines and the base of length b. Before overlapping these lengths were

$$\tfrac{1}{5}b, \quad \tfrac{2}{5}b, \quad \tfrac{3}{5}b, \quad \tfrac{4}{5}b, \quad b.$$

Afterwards 4 overlaps, each of $b/20$, occur at all levels but the first, and the totals of intercepts become

$$\tfrac{1}{5}b, \quad \tfrac{1}{5}b, \quad \tfrac{2}{5}b, \quad \tfrac{3}{5}b, \quad \tfrac{4}{5}b.$$

The next stage overlaps 2 sets of pairs at the third level to produce Figure 18. Total lengths along guide lines and base are now

$$\tfrac{1}{5}b, \quad \tfrac{1}{5}b, \quad \tfrac{1}{5}b, \quad \tfrac{2}{5}b, \quad \tfrac{3}{5}b,$$

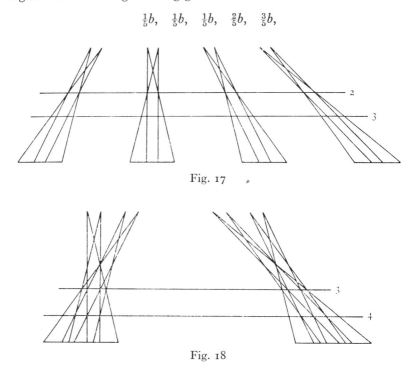

Fig. 17

Fig. 18

as $\tfrac{1}{5}b$ is subtracted from the last 3 entries. Finally we overlap the 2 sets of 4 triangles on the fourth guide line obtaining Figure 19. Total lengths intercepted are now $\tfrac{1}{5}b$ on all but the base, where $\tfrac{2}{5}b$ remains. From these totals the area of the figure is readily found to be $\tfrac{1}{5}bh$, where h is the height of the triangle. So the original area has been multiplied by $\tfrac{2}{5}$. For general m this factor becomes $2/(m+2)$, and can be made as small as we wish. Note that in the overlap process only parallel translations of triangles have occurred.

We could rotate a rod of length less than h from AB to AC, sweeping out an area less than that of the triangle, but still of appreciable size. Instead we achieve the same final result by successive rotations through the subtriangles alternating with the translations they suffer in the

98

overlap process. The total area swept out in rotation will then be less than $2/(m+2)$ times that of the original triangle, for it is less than the total overlapped area. There are $(2^m - 1)$ parallel shifts, and the method outlined earlier enables these to be done so that the total area covered is less than

$$bh/(m+2)+(2^m-1)\,2l^2\phi.$$

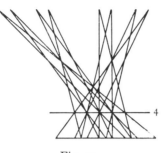

Fig. 19

Choose m large enough to make the first term as small as required, and then choose ϕ to make the second equal to it. We have thus turned the rod through angle BAC in as small an area as we care to specify. By making this angle $\frac{1}{2}\pi$ and repeating the process, a rotation through π in an arbitrarily small area is achieved. Of course very many moves are involved, combined with motions of the rod along its length for enormous distances.

References

1. V. G. Boltyanskii. *Equivalent and Equidecomposable Figures* (D. C. Heath and Co., 1963).
2. J. H. Cadwell. 'Some Dissection Problems involving Sums of Cubes.' *Math. Gaz.* **48**, no. 366, 391 (1964).
3. A. Ehrenfeucht. *The Cube made Interesting*. (Pergamon Press, 1964).
4. M. Gardner. *More Mathematical Puzzles and Diversions* (Bell, 1963).
5. M. Kraitchik. *Mathematical Recreations* (George Allen and Unwin, 1943).
6. H. Lindgren. *Geometric Dissections* (van Nostrand, 1964).
7. W. W. Rouse Ball. *Mathematical Recreations and Essays*, 11th edition (Macmillan, 1940).
8. S. K. Stein. *Mathematics the Man-made Universe* (W. H. Freeman and Co., 1963).
9. H. Steinhaus. *Mathematical Snapshots* (Oxford, 1960).
10. D'Arcy W. Thompson. *On Growth and Form*, abridged edition (Bonner) (Cambridge, 1961).
11. H. Meschkowski. *Unsolved and Unsolvable Problems in Geometry* (Oliver and Boyd, 1966).

10

Newton's polygon and plane algebraic curves

1. *The Folium of Descartes*

Descartes, the discoverer of coordinate geometry, studied the curve of Figure 1 (*a*). Its equation is

$$x^3 + y^3 = 3axy, \qquad (1)$$

and we see that neither x nor y can be expressed explicitly in terms of the other. Newton carried out a classification of cubics, and devised a powerful method for plotting curves. Before applying it to (1) we shall make a few general comments on curve tracing.

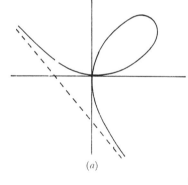

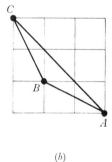

(*a*)   (*b*)

Fig. 1

The general equation of a straight line is

$$Ax + By + C = 0,$$

and if we substitute for y in (1) we get a cubic equation in x indicating that there are 3 points of intersection. However, if the line touches the curve, or goes through its double point, 2 of the roots will coincide. Still more special is an inflectional tangent which both touches and crosses a curve, in this case 3 roots coincide at the point of contact. We see later that the Folium has such a tangent. We may lose 2 roots too if our line, besides going through a double point, touches the curve there. It is a general result that, for a curve through the origin, the tangent(s) there come from the homogenous group of terms of lowest degree in its equation. Here this group consists of the right-hand side of (1), so the axes $x = 0$, $y = 0$ touch the curve at the origin.

Some lines lose points of intersection with a curve in quite a different way. Thus the line

$$x+y+C = 0,$$

when combined with (1), leads to a quadratic. The vanishing of the cubic term indicates that, as C varies, these lines all go through a point at infinity on the curve. One of them, with $C = a$, when combined with (1) leads to the result $a^3 = 0$, indicating that all 3 roots are at infinity.

This line, shown dotted in Figure 1(a), is an asymptote; it touches the curve at infinity. It is in fact a rather special asymptote, for the curve has an inflection at the infinite point of contact. This accounts for the loss of 3 roots rather than 2, and for the fact that the curve comes in from infinity on the same side of the line at either end.

Newton's polygon ABC is shown in Figure 1(b). The points A, B and C arise from the 3 terms of (1) and are placed on a square mesh as follows. The term in x^3 gives A with coordinates (3, 0) on this mesh, y^3 gives C or (0, 3), while xy gives B or (1, 1). Thus it is the indices of x and y that determine points on the mesh. Each side of the polygon indicates a branch of the curve. We now show how these are used, leaving proofs until the next section.

Side AB has no marked points below it, and gives a branch at the origin. The terms corresponding to A, B are retained from (1) giving

$$x^3 - 3axy = 0 \quad \text{or} \quad x^2 = 3ay,$$

the equation of a parabola touching $y = 0$. Similarly, side BC leads to the branch

$$3ax = y^2,$$

a parabola touching the y axis. Figure 2 indicates these branches in sketches (a) and (b). The Folium resembles these curves along their fully marked portions; we assume that $a > 0$.

The side CA has only points below it and slopes at $135°$, indicating linear asymptotes. Their directions are given by

$$x^3+y^3 \equiv (x+y)(x+\omega y)(x+\omega^2 y) = 0,$$

101

and only one indicated a real line. To find the corresponding asymptote we include all 3 terms, and write (1) as

$$x+y = \frac{3axy}{x^2-xy+y^2} \simeq \frac{-3ax^2}{3x^2} = -a. \qquad (2)$$

The approximate result follows by substituting $y = -x$, and the asymptote is $x+y+a = 0$.

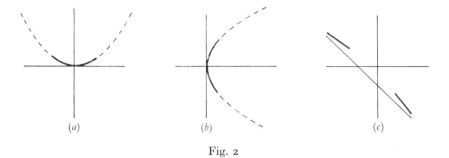

(a) (b) (c)

Fig. 2

It is useful to take this approximation one stage further. This time y is replaced by $-(x+a)$ giving

$$x+y = \frac{3axy}{x^2-xy+y^2} \simeq \frac{-3ax(x+a)}{3x^2+3ax+a^2} \simeq -a+\frac{a^3}{3x^2}.$$

The final result is obtained by simple division. The term in $1/x^2$ indicates that the curve lies above the asymptote for both large positive and negative x, as in Figure 2(c).

There are some general questions we ask before proceeding to join up the various branches. Does the curve cut either axis (apart from doing so at the origin)? Does the asymptote have any finite intersections with the curve? The negative answers to these questions help by reducing possible modes of joining branches, leaving only the one shown.

Had there been oblique tangents at the origin, their points of intersection with the curve would also supply useful information. As interchange of x and y leaves (1) unaltered there is symmetry about the line $y = x$. It is useful to determine the point for which $y = x$, apart from the origin.

Whenever a curve has a multiple point at the origin it is worth seeing if a parametric representation can be found. The line $y = mx$ will have several of its intersections with the curve at $(0, 0)$; if, as here, only one

102

other remains, the curve can be expressed parametrically. The Folium can be defined by

$$x = \frac{3am}{1+m^3}, \quad y = \frac{3am^2}{1+m^3},$$

a result that makes accurate plotting easy.

2. *Newton's polygon*

Having plotted the indices of all terms of a curve's equation on a square mesh there will be a unique convex polygon joining some of the marked points, and including the rest within it. Let PQ be one of its sides such that all other marked points lie on the opposite side of PQ to the origin. The point R is typical of these other points, the corresponding terms in the equation being

$$\cdot \qquad P \equiv ax^\alpha y^\beta, \quad Q \equiv bx^\gamma y^\delta, \quad R \equiv cx^\xi y^\eta.$$

The line PQ will have negative slope, so that, if $\alpha > \gamma$ then $\delta > \beta$. After dividing out a common factor, the 3 terms P, Q and R give

$$ax^{\alpha-\gamma} + by^{\delta-\beta} + cx^{\xi-\gamma}y^{\eta-\beta}.$$

The approximation (4) obtained by setting the sum of the first 2 terms equal to zero can be used to eliminate y in the third. The index of x after this step will be

$$\xi - \gamma + (\eta - \beta)\frac{\alpha-\gamma}{\delta-\beta}.$$

This exceeds the index of x in the first term by

$$\xi - \alpha + (\eta - \beta)\frac{\alpha-\gamma}{\delta-\beta}, \tag{3}$$

and if (3) is positive, the third term, involving a higher power of x than the first, can be neglected for small values of x.

Introducing coordinates (X, Y) in the polygon diagram, the equation of PQ is

$$(\delta-\beta)(X-\alpha) + (\alpha-\gamma)(Y-\beta) = 0.$$

The perpendicular distance from R to this line is

$$\{(\delta-\beta)(\xi-\alpha) + (\alpha-\gamma)(\eta-\beta)\}/\sqrt{\{(\delta-\beta)^2 + (\alpha-\gamma)^2\}},$$

a positive quantity, as R and the origin lie on opposite sides of the line. Thus (3) is positive, and we can approximate the curve by

$$ax^{\alpha-\gamma} + by^{\delta-\beta} = 0, \tag{4}$$

103

for points near the origin. By moving PQ parallel to itself until it meets a marked point we determine which other term should be included to approximate one stage further. Note that, if the curve does not pass through the origin, no branches of this sort occur in the polygon.

A precisely similar argument shows that a side of the polygon with all marked points lying below it gives an approximation valid for large x and y. If this line slopes at $135°$, the corresponding terms will be of the same total degree in x and y, and, in general, linear asymptotes result. Sides at other slopes lead to curvilinear asymptotes. The way in which asymptotes parallel to the axes are indicated will appear in one of the examples.

3. *A curvilinear asymptote*

We consider the curve

$$4x^2 - 2xy^2 + y^3 = 0,$$

the Newton polygon, again a triangle, is shown in Figure 3(b). The side AB corresponds to

$$4x^2 + y^3 = 0,$$

an approximation near the origin leading to Figure 4(a).

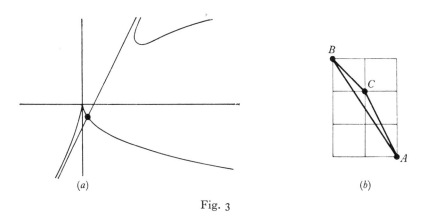

(a) (b)

Fig. 3

The side BC at $135°$ indicates a linear asymptote; writing the dominant terms B and C first, we have

$$y^2(y - 2x) + 4x^2 = 0.$$

Putting $y = 2x$ everywhere except in the bracket leads to the asymptote

$y = 2x - 1$. Replacing y by $(2x-1)$ and using the binomial expansion or simple division gives the better approximation

$$y = 2x - 1 - \frac{1}{x}.$$

This implies that the curve lies below the line for positive x, and vice versa; Figure 4(b) illustrates this result. We note that, in the general case, before substituting a better value for y another term besides A would be included with B and C. This is selected by further parallel displacement of BC. Here there are no other terms left to consider.

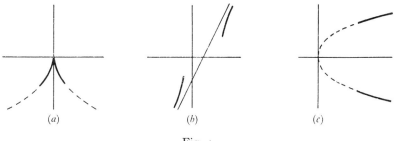

(a)   (b)   (c)

Fig. 4

The side CA indicates another infinite branch,

$$4x^2 - 2xy^2 = 0 \quad \text{or} \quad y^2 = 2x,$$

Figure 4(c) shows this parabolic asymptote.

The curve does not cut the axes other than at $(0, 0)$. To find its intersection with the linear asymptote we put $y = 2x - 1$, to get

$$0.x^3 + 0.x^2 + 4x - 1 = 0,$$

indicating, as it should, two roots at infinity, together with the finite point $(\frac{1}{4}, -\frac{1}{2})$. It is now easy to sketch the curve.

4. *Asymptotes parallel to the axes*

Our next example

$$2x - x^2 y - xy^2 - y^4 + xy^4 = 0,$$

gives the polygon of Figure 5(b).

The branch AB is

$$2x = y^4,$$

105

and supplies Figure 6(a). The horizontal side BC gives a vertical asymptote

$$xy^4 - y^4 \equiv y^4(x-1) = 0,$$

or $x = 1$. Moving BC parallel to itself we first encounter the point corresponding to xy^2, so that a better approximation is

$$y^4(x-1) - xy^2 = 0 \quad \text{or} \quad x = 1 + (1/y^2),$$

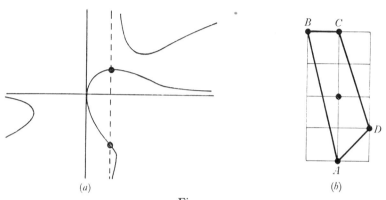

(a) (b)

Fig. 5

obtained by putting $x = 1$ except in the bracket itself. Thus, as shown in the sketch in Figure 6(b), the curve lies to the right of $x = 1$ for both positive and negative y.

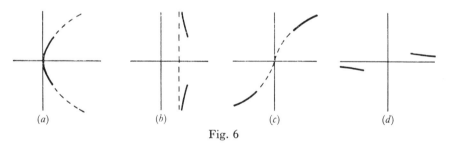

(a) (b) (c) (d)

Fig. 6

Side CD gives the curvilinear asymptote

$$x = y^3$$

of Figure 6(c). Branch DA has a positive slope, and the horizontal axis itself is an asymptote, the terms D and A give

$$2x - x^2y = 0 \quad \text{or} \quad y = 2/x$$

so that the curve lies above the axis for $x > 0$, and vice versa, as in Figure 6(d).

106

The line $x = 1$ cuts the curve at the finite points with $y = -2, 1$. There is still some uncertainty left as to how the various branches should be joined. We note that the line $y = 1$ cuts the curve twice at $x = 1$, and at no other point. We can now produce a sketch of the curve.

5. *Oblique tangents at the origin*

We consider
$$(y-x)^2(y+x) = x^4 + 4y^4, \qquad (5)$$

sketched in Figure 7(a). Side AB indicates an intersection with the y axis at distance $\frac{1}{4}$; CD leads to the point $(1, 0)$ on the curve. The only infinite branch is
$$x^4 + 4y^4 = 0.$$

This has no real factor, so the curve is of finite extent.

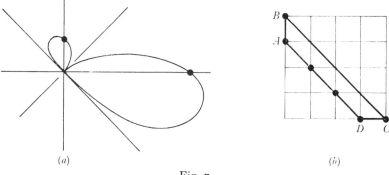

(a) (b)

Fig. 7

The side DA includes four terms, which together give
$$(y-x)^2(y+x) = 0,$$

indicating oblique tangents $y = \pm x$ at $(0, 0)$. We note that a segment sloping at $135°$ always leads to oblique tangents. Parallel translation of DA brings in all terms of (5). Putting $y = -x$ everywhere except in the critical bracket we get
$$y+x = \tfrac{5}{4}x^2,$$

so the curve lies above this tangent on either side of zero.

Putting $y = x$ except in its own bracket leads to
$$(y-x)^2 = \tfrac{5}{2}x^3 \quad \text{or} \quad y = x \pm \sqrt{\tfrac{5}{2}}\, x^{\frac{3}{2}}.$$

107

This indicates a cusp, for there are no points near the origin for negative x. For $x > 0$ there are pairs of points one on each side of the tangent.

We note that neither tangent cuts the curve at points other than the origin, and a sketch is readily prepared. The intersection of the curve and the line $y = mx$ again leads to a parametric form.

6. A rhamphoid cusp

We consider the curve
$$(y-x^2)^2 = xy^2 - y^4, \tag{6}$$

sketched in Figure 8(a). Side BC of the polygon gives zero as the only real intersection with $x = 0$. The side CA leads to
$$x^4 + y^4 = 0,$$

without real factors, and therefore indicating a finite curve.

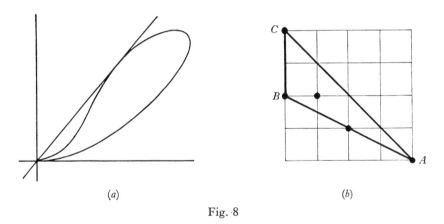

(a)   (b)

Fig. 8

The side AB contains three terms giving
$$(y-x^2)^2 = 0.$$

Had this been the product of different factors it would lead to two parabolic branches touching the x axis. As it is we must go to the next stage
$$(y-x^2)^2 = xy^2,$$

and replace the y on the right-hand side by x^2, getting
$$y = x^2 \pm x^{\frac{5}{2}}.$$

108

For $x < 0$ there are no real points, for $x > 0$ they occur in pairs one on each side of $y = x^2$, leading to a rhamphoid cusp. The simpler cusp of the last example, with the two branches of the curve on opposite sides of the tangent, is called ceratoid.

The line $y = mx$ cuts the curve in two points given by

$$x^2(1+m^4) - x(2m+m^2) + m^2 = 0, \tag{7}$$

a quadratic with real roots for values of m lying between 0 and the real root of the cubic

$$4m^3 - m - 4 = 0,$$

obtained by equating the discriminant of (7) to zero. This root is close to 1·08, and indicates the tangent from the origin to the curve shown in the figure. As a further guide to sketching, the quadratic can be solved for $m = 0·5$. Here, although the curve is not rational, i.e. does not admit of a parametric representation, its intersections with lines through a multiple point still provide valuable information for plotting purposes.

7. A singularity at infinity

The curve

$$y - 2x + x(y-x)^2 = 0, \tag{8}$$

has a single tangent at the origin. The improved approximation to the curve there is

$$y = 2x - x^3,$$

so that the inflectional tangent is crossed by the curve at zero.

The side BC of the polygon in Figure 9(b) is of positive slope and indicates that the vertical axis is an asymptote, the two terms give

$$y + xy^2 = 0 \quad \text{or} \quad x = -1/y,$$

the curve is to the left of the axis for $y > 0$, and vice versa.

Side CD indicates oblique asymptotes, and as the direction is squared we expect them to be parallel. The next approximation is (8) itself, putting $y = x$ except in the critical bracket gives

$$x(y-x)^2 - x = 0 \quad \text{or} \quad y = x \pm 1.$$

We improve on this by putting $y = x+1$, noting that there are no further terms to include at this stage. The result is

$$y = x + \sqrt{\{1 - (1/x)\}}$$

so that the curve lies above the line for $x < 0$ and vice versa. For the other asymptote the situation is similar.

The side AD indicates points $(\pm\sqrt{2}, 0)$ on the curve. There are no non-zero intersections with $y = 2x$, the tangent at the origin; or finite intersections with any of the asymptotes. Changing the sign of both x and y in (8) leaves it unaltered, so the curve has central symmetry about the origin.

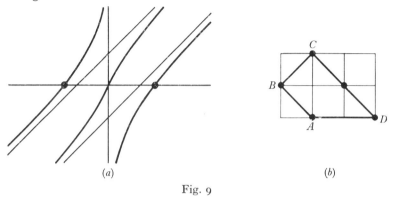

(a)   (b)

Fig. 9

The parallel asymptotes arise from a double point of the curve at infinity. In our last example we shall meet a squared asymptotic direction factor with a quite different result.

8. *Contact with the line at infinity*

We consider
$$x^2 - y^3 + (x^2 - y^2)^2 = 0.$$

Behaviour at the origin corresponds to BC, so there is a simple cusp there. Sides AB and CD lead to one real axial point $(0, 1)$.

Side AD has three points on it, leading to
$$(y - x)^2(y + x)^2 = 0.$$

Parallel translation shows that C must next be included to give
$$(y - x)^2(y + x)^2 - y^3 = 0.$$

Putting $y = x$ except in the critical bracket gives
$$4(y - x)^2 = x.$$

110

This time the squared term leads, not to parallel asymptotes, but to an asymptotic parabola, with its axis parallel to $y = x$. Similarly the other squared factor gives

$$4(y+x)^2 + x = 0.$$

These parabolas have been sketched as guide curves in Figure 10(a).

Whereas in Section 7 the repeated factor indicates a double point at infinity, here it arises because the curve touches the line at infinity. In fact it does so twice, and at each point can be replaced by an approximating parabola. The curvilinear asymptotes we encountered previously arose because the curves involved touched the line at infinity at either $x = 0$ or $y = 0$.

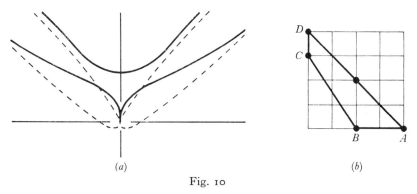

(a)   (b)

Fig. 10

Newton's polygon seems to be little known, although Frost[2] gives a large number of examples of its use. In addition, Walker[4], besides discussing algebraic plane curves systematically, includes a section on this topic. We mention too the interesting introduction to algebraic curves by Forder[1]. Although not directly related to the present topic, fascinating discussions of plane curves both algebraic and transcendental, are to be found in Lockwood,[3] Yates[5] and Zwikker[6].

References

1. H. G. Forder. *Geometry* (Hutchinson's Home University Library, 1960).
2. P. Frost. *Curve Tracing* (Chelsea Publishing Co., 1960).
3. E. H. Lockwood. *A Book of Curves* (Cambridge, 1961).
4. R. J. Walker. *Algebraic Curves* (Dover, 1962).
5. R. C. Yates. *Curves and their Properties* (Michigan, 1947).
6. C. Zwikker. *The Advanced Geometry of Plane Curves and their Applications* (Dover, 1963).

11

The plane symmetry groups

1. *Symmetry and isometric motion*

The patterns shown in Figures 11 to 16 of this chapter (pp. 123–128) are all possessors of obvious regularity or symmetry. At the same time it is evident that the type of symmetry varies from pattern to pattern. We shall prove that the 7 one-dimensional patterns of Figure 11 include all possibilities of this kind. There are 2 possible types of point symmetry, these are shown in Figure 8; each type has a single infinity of variants. Of two-dimensional patterns there are just 17 types. We shall not enumerate these in detail; however, they are all illustrated in Figures 12 to 16.

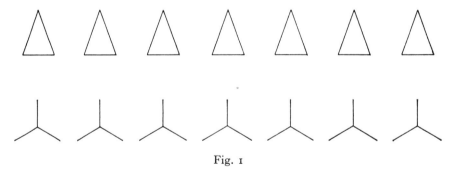

Fig. 1

We start by defining symmetry in terms of rigid or isometric motions of the plane. It is evident that both one- and two-dimensional schemes can be moved in various ways until the pattern in its new position lies over the old. Besides translations, this can be achieved in some cases by suitably rotating the plane. Further, some patterns are invariant when reflected in certain lines. We say that a pattern is characterised by the totality of isometric motions that bring it into self-coincidence.

In connection with this definition of symmetry we note that the two patterns of Figure 1 belong to the same class. While rotations through $\frac{2}{3}\pi$ leave the pattern elements of the second unaltered, they do displace the pattern as a whole, and from our point of view they add nothing to the symmetry properties of the first diagram.

An isometric transformation or rigid motion of the plane preserves distances and angles. There are two distinct types, illustrated in Figure 2. Diagrams (*a*) and (*b*) show the same sense of rotation in their lettering, and (*a*) can be laid exactly over (*b*) by a suitably chosen isometry, as we shall see in the next section. However, (*c*) has an opposite rotation, and no rigid motion of (*a*) in its own plane can secure coincidence with it. It is possible to rotate (*a*) into the dotted triangle. Then a further change, a reflection in $A''B''$, turns this into (*c*). We can regard this last change as a rigid motion of the plane obtained by rotating it about the line $A''B''$ through angle π. Isometries obtained by motions lying entirely in the plane are called proper. Those requiring a rotation of the plane out of itself are called improper.

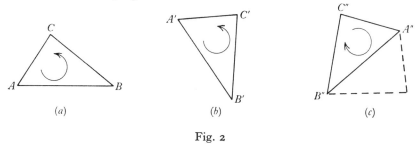

(*a*)   (*b*)   (*c*)

Fig. 2

A fundamental property of isometries is that any two carried out in succession define a third. Thus if isometry **U** takes figure F into figure F', while **V** takes F' into F'', their combined effect, taken in this order, is an isometry carrying F into F''. We denote this 'product' isometry by the symbol **VU**, our convention being that operation **U** is applied before **V**. As a proper isometry **P** does not affect rotational sense, while an improper isometry **I** alters it, we see that

> P_1P_2 is proper,
> **PI** or **IP** is improper,
> I_1I_2 is proper.

2. *Rotations and glide reflections*

We first prove that any proper isometry is equivalent to a rotation. Let the isometry move triangle ABC into $A'B'C'$. In Figure 3 (*a*) we join AA' and BB' and construct their perpendicular bisectors meeting in O.

Then, as triangles AOB, $A'OB'$ have equal sides, they are congruent. By adding angle BOA' to their angles at O we see that angles AOA' and BOB' are equal. A rotation through this angle α about O thus carries AB into $A'B'$. By adding angle OAB to BAC and $OA'B'$ to $B'A'C'$, and noting that $OA = OA'$, $AC = A'C'$, we prove that triangles OAC, $OA'C'$ are congruent. It follows that $OC = OC'$ and that angle COC' is also equal to α. Hence the same rotation carries point C into C'.

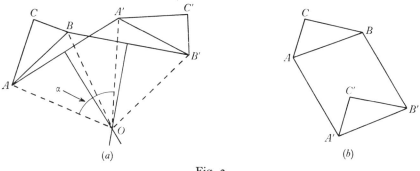

Fig. 3

The above construction fails if the perpendicular bisectors are parallel, as in Figure 3 (b). We agree to call this translatory motion a rotation through zero angle about a suitable point at infinity. This then justifies our claim that any proper isometry is a rotation.

It is easy to construct an example to show that the products S_1S_2 and S_2S_1 of 2 rotations or spins S_1 and S_2 differ, or we can prove this as follows. Let O_1, O_2 be the centres of rotation, and consider the effects of the 2 products on O_1. We note that S_1 leaves O_1 unchanged, or

$$S_1(O_1) = O_1.$$

Hence
$$S_2S_1(O_1) = S_2(O_1),$$

$$S_1S_2(O_1) = S_1(S_2(O_1)),$$

and the two mappings of O_1 differ, for S_1 displaces all points except O_1, in particular it moves $S_2(O_1)$. It is useful to note that, for either product, the resultant angle of rotation is the sum of the two constituent angles.

We define a basic improper isometry called a glide reflection. In Figure 4, triangle ABC is reflected in line l, and then moved parallel to it through distance d. The improper motion taking ABC into $A'B'C'$ is a glide reflection defined by l (with a direction along it) and d.

114

Any improper isometry is obtainable by a glide reflection. The direction of l bisects the angle between AB and $A'B'$, and it passes through the mid-point of AA'. The point in which a line through A', parallel to l, cuts the circle on AA' as diameter determines d. So the glide reflection needed to take ABC into $A'B'C'$ is completely specified.

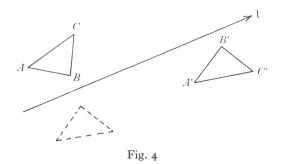

Fig. 4

3. *Groups of isometries*

A group of operations is a set possessing the following properties:

(i) The application of two operations in sequence produces the same result as some other operation of the group. In symbols, if the result of $\mathbf{U}$ followed by $\mathbf{V}$ is another element $\mathbf{W}$, we write $\mathbf{W} = \mathbf{VU}$.

(ii) There is an identity operation $\mathbf{I}$ with the properties

$$\mathbf{UI} = \mathbf{IU} = \mathbf{U}$$

for all members $\mathbf{U}$ of the group.

(iii) Each member $\mathbf{U}$ has an inverse called $\mathbf{U}^{-1}$ in the group with the properties $\mathbf{UU}^{-1} = \mathbf{U}^{-1}\mathbf{U} = \mathbf{I}$.

(iv) Operations combine associatively, so that $\mathbf{U(VW)} = \mathbf{(UV)W}$.

We first consider all plane translations. Any two translations can be combined by the parallelogram law to give a resultant translation, so that (i) is satisfied. It is also true here that $\mathbf{T_1T_2} = \mathbf{T_2T_1}$, so we say the operators are commutative. There is a translation through zero distance that we call $\mathbf{I}$, and if $\mathbf{T}$ is a translation through distance d, then $\mathbf{T}^{-1}$ is a parallel translation through $-d$. The result (iv) holds, so that we may speak of the group of plane translations.

In an exactly similar way all rotations form a group; this includes translations as special cases. We say that the translation group is a subgroup of the group of plane rotations. We note that the latter group is not commutative. Since the product of a pair of improper isometries is proper, the improper motions do not form a group. Adding them to the rotation group we arrive at all possible isometric operations, the group of plane isometries.

All the above groups are continuous. This means that we can find operations that will map a point P as near as we care to specify to any point of the plane. The symmetry groups are discrete, and although a point P will have an infinite number of mappings, no two of these are closer together than some finite distance.

4. *Examples of discrete groups*

We consider the first diagram of Figure 11. Here there is a basic translation **T** through distance d, taking the pattern along through one of its elements. The infinite set of operations forming the group consist of all powers of **T**, i.e. of all shifts of a positive or negative multiple of d. They are

$$\ldots, \ \mathbf{T^{-2}}, \ \mathbf{T^{-1}}, \ \mathbf{I}, \ \mathbf{T}, \ \mathbf{T^2}, \ \mathbf{T^3}, \ \ldots \tag{1}$$

Next we consider the third diagram. Besides the operations of (1) there are reflections in a series of axes at spacing $\frac{1}{2}d$. If **R** denotes reflection in any one of these axes, Figure 5 shows that $\mathbf{T}^k\mathbf{R}$ is a reflection

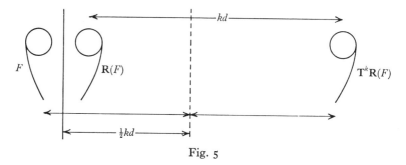

Fig. 5

in an axis at distance $\frac{1}{2}kd$ from that of **R**. This is another axis of the system. Thus by premultiplying **R** by the sequence (1) we generate all reflections of the group. We say that (1) is a subgroup of this new group; by combining (1) with any member outside itself all others are generated. These others, being improper, do not form a subgroup. We refer to them as the coset corresponding to subgroup (1).

We note that any member of (1) when applied to a reflection axis produces another such axis. Similarly, if a discrete group possesses a

116

centre of rotation; translations, rotations and reflections of the group will generate further centres when applied to it. We shall frequently use this property in the process of enumeration.

A pattern is characterised by the set of isometric motions that bring it into self-coincidence, and this set forms a discrete group. Thus the enumeration of patterns is equivalent to the enumeration of discrete groups of isometric motions in the plane.

5. *Similarity transformations*

The theory of plane isometries is developed by Yaglom[4], who describes some fascinating applications to geometrical problems. We content ourselves with examining the idea of a similarity transformation applied to an isometry. This concept proves very useful as a tool in enumerative proofs. Let $\mathbf{T}$ be a translation through distance d in direction l, and $\mathbf{S}$ be a rotation. We consider the sequence of three operations denoted by $\mathbf{STS^{-1}}$, and prove that it is also a translation through d, but parallel to a line $\mathbf{S}(l)$ obtained by rotating l by $\mathbf{S}$.

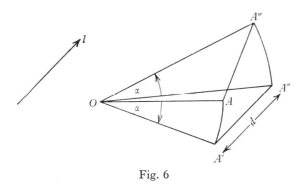

Fig. 6

The proof follows from Figure 6, where $\mathbf{S^{-1}}(A)$ gives A', by a negative rotation of magnitude α about O. Translation $\mathbf{T}$ moves A' through d parallel to l giving A''. Then $\mathbf{S}(A'')$ gives A''' by the rotation through α taken positively. The result, A to A''', is a translation through d, and its direction makes angle α with l, i.e. it is l's direction after modification by $\mathbf{S}$.

It is left to the reader to verify that, for rotations $\mathbf{S_1}$ and $\mathbf{S_2}$, the product $\mathbf{S_2 S_1 S_2^{-1}}$ is a rotation whose magnitude is that of $\mathbf{S_1}$, and whose centre is $S_2(O_1)$, the image of O_1 by rotation S_2. The product $\mathbf{RTR^{-1}}$ is a translation whose direction is obtained by applying reflection $\mathbf{R}$ to that of $\mathbf{T}$.

6. *The point groups*

The plane symmetry groups are discrete, and we base our enumeration upon their translational structure. There are three cases to consider:

(i) No translations present, the point groups.

(ii) Only one direction of translation, the frieze groups.

(iii) More than one direction of translation, the wallpaper groups.

The first of these is the subject of this section. Such a group can contain no glide reflections, for the square of a glide reflection is readily seen to be a translation. If there is more than one reflection, the group contains their product, a rotation. The case of a single reflection is shown in Figure 7(a). The group contains two operators, $\mathbf{I}$ and $\mathbf{R}$, with $\mathbf{R^2 = I}$. We know that all others have rotations, and our first step will be to prove that there can be only one centre of rotation.

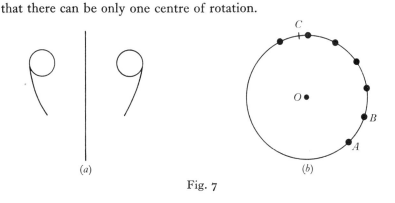

(a)   (b)

Fig. 7

Suppose there are two rotations or spins S_1 and S_2 with different centres O_1 and O_2, and of magnitudes α_1 and α_2. We have already said that the operation

$$S = S_2 S_1 S_2^{-1}$$

is a rotation through α_1 about the point $S_2(O_1)$. Now we consider $S_1^{-1}S$; this is a rotation of magnitude $(-\alpha_1+\alpha_1)$ or zero, i.e. it is a translation. By (i) it must be the zero translation so that

$$S_1^{-1}S = I \quad \text{or} \quad S = S_1.$$

The last result, obtained by multiplying both sides of the first expression by S_1, implies that S and S_1 have the same centre or that $S_2(O_1) \equiv O_1$,

118

a contradiction. Thus we are forced to drop the assumption of more than one centre of rotation.

Our next step is to prove that all rotations about the one centre are by multiples of an angle $2\pi/n$, where n is an integer greater than 1. The point A of Figure 7(b) has among its mappings by the group one of minimum separation from A itself, for our group is discrete. Call it B, and let C be any other mapping of A. Let angle $AOB = 2\pi/n$, and the corresponding isometry be $\mathbf{S}$. We wish to prove that AOC is an integer multiple of this angle. If not, then we can find an integer k so that

$$\frac{2\pi k}{n} < \widehat{AOC} < \frac{2\pi(k+1)}{n}.$$

The operation $\mathbf{S}^{-k}$ brings C to a point inside the smaller arc AB, contradicting our definition of B. Thus all images of A under the group consist of an equally spaced set of points. One of the multiples of $2\pi/n$ must be

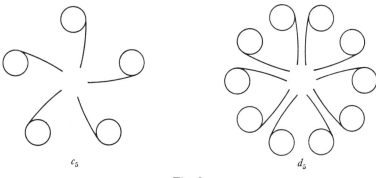

c_5 d_5

Fig. 8

2π, for A comes back to its old position after one revolution, and n must be an integer. The first diagram of Figure 8 illustrates the point group c_n for $n = 5$.

We must now see how reflections can be added. If an axis of reflection did not go through O, then the reflection of O in this axis would furnish a second centre of rotation. This is inadmissible, so any axes of reflection pass through O. Any one axis will provide $(n-1)$ others under the rotational operations of the group. A reflection $\mathbf{R}$, followed by a rotation $2\pi/n$ about a point on the reflecting axis, is easily seen to be equivalent to a reflection in an axis at π/n to the first one. This gives a second set of axes bisecting the first.

Next we observe that the product of any two reflections is a rotation through twice the angle between their axes (see Figure 9(a)). So any two

119

axes of reflection of the group must be separated by half a multiple of $2\pi/n$. This means that all such reflections have been accounted for. We have the point group d_n with $2n$ axes of reflection, illustrated above for $n = 5$. If we consider c_1 as consisting of the identity operation only, and d_1 to correspond to Figure 7(a), then c_n and d_n include all point groups. Leonardo da Vinci is said to have enumerated these groups.

7. The frieze groups

These correspond to (ii) at the beginning of Section 6, involving a single direction of translation. It is readily found that the translations are all integer multiples of a basic distance d; this follows closely the similar investigation for rotations in the point groups. The fundamental pattern is thus the first of Figure 11 and it is called F_1. We now examine what rotations can be added. If there is a rotation $\mathbf{S}$, the translation $\mathbf{STS^{-1}}$ obtained by transforming $\mathbf{T}$ by this operation must be parallel to

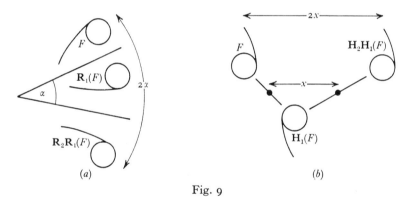

Fig. 9

$\mathbf{T}$, for only one direction of translation is allowed. Thus $\mathbf{S}$ must be of magnitude 0 or π, i.e. only half-turns can be admitted. We derived d_n from c_n by introducing reflections. The introduction of half-turns into F_1 gives F_2 in a very similar manner. Figure 9(b) shows that the product of two half-turns is a translation through twice the distance between their centres. As $2x = kd$ then $x = k.\frac{1}{2}d$, and the half-turn centres at a spacing of $\frac{1}{2}d$ are directly comparable with the set of reflection axes in d_n.

We now admit reflections, and the similarity transformation $\mathbf{RTR^{-1}}$ shows that the direction of $\mathbf{T}$ must be unaltered by any $\mathbf{R}$, just as it was by any $\mathbf{S}$. Therefore reflection axes must be parallel to $\mathbf{T}$, or perpendicular

120

to it. A pair of parallel reflecting axes generate a translation through twice the distance between them. From this fact we make two deductions. First, there can only be one reflection axis parallel to $\mathbf{T}$, for two would generate a translation perpendicular to it. Secondly, if there is a reflection axis perpendicular to $\mathbf{T}$, there will be an array at spacing $\frac{1}{2}d$. Again, the arguments that show the existence and completeness of this array are similar to those used in moving from c_n to d_n, or in introducing half-turns into F_1. A reflection, followed by a perpendicular translation d, gives a reflection in an axis moved through $\frac{1}{2}d$. Thus there are a pair of axes $\frac{1}{2}d$ apart. Then $\mathbf{T}$ and its powers operating on these generate the whole array. A pair of parallel reflections is equivalent to a translation through twice the distance between their axes. So any two reflections must be separated by half a multiple of d, the group's translation distance. Therefore all reflections are included in the above array.

An axis of reflection introduced parallel to $\mathbf{T}$ in F_1 gives F_1^1, and a set of axes perpendicular to it results in F_1^2. It is easy to see that, if both types of axes occur, their intersections become half-turn centres, so we really get a derivative of F_2 called F_2^1. We must consider the introduction of a parallel axis alone in F_2. Such an axis would have to pass through the half-turn centres, or another line of these centres would result by reflection in it. It is then readily seen that the product of a reflection and a half-turn would be equivalent to a reflection in an axis perpendicular to $\mathbf{T}$ so we are back to F_2^1. Finally, we consider an array of perpendicular axes only. This array must either pass through the array of half-turn centres or bisect this array. Otherwise reflections of these centres would produce a new set. The first case again implies reflection in the line of centres, i.e. it leads to F_2^1. The second gives a new scheme, that of F_2^2.

To round off the enumeration the possibility of glide reflections $\mathbf{G}$ must be considered. It is not difficult to show that only one axis, and that one parallel to $\mathbf{T}$, is possible. Further, half-turn centres, if present, must lie upon it. Now $\mathbf{G}^2$ is a translation, and therefore

$$\mathbf{G}^2 = \mathbf{T}^{2k} \quad \text{or} \quad \mathbf{G}^2 = \mathbf{T}^{2k+1}.$$

In the first case $\mathbf{T}^{-k}\mathbf{G}$ is a glide reflection whose square is the identity Thus it is a simple reflection, and $\mathbf{G}$ can add nothing to the patterns already found, for introducing this reflection is equivalent to including $\mathbf{G}$. In the second case $\mathbf{T}^{-k}\mathbf{G}$ is a glide through $\frac{1}{2}d$. We omit details, but the possibility of combining this operation with the six existing patterns leads to a single new one called F_1^3. In its first pattern unit the result of the reflection part of the glide has been dotted in.

8. The wallpaper groups

We shall not attempt the enumeration of the cases under (iii) of Section 6. The first step is to prove that there are two directions of translation in terms of which all others may be expressed. The method is an extension of that used to show that only one basic rotation entered with the point groups. The resultant simplest wallpaper pattern is W_1 of Figure 12.

We then enquire what rotations can be allowed; we shall prove that only the angles $2\pi/n$ need be considered for $n = 2, 3, 4$ and 6. The corresponding basic patterns are the first ones in Figures 13 to 16.

In Figure 10 let P be a centre of rotation, angle $2\pi/n$, and let Q be the nearest centre to it of the same type. Then rotating the pattern through $2\pi/n$ about Q must produce another centre of this type at R. Rotation about R will put yet another at S. If n is greater than 6, PR will be less than PQ, a contradiction of our assumption that Q is closer to P than any other '$2\pi/n$' centre. With $n = 6$, S and P coin-

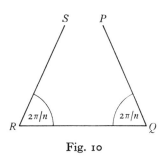

Fig. 10

cide; $n = 5$ makes PS less than PQ so it is not a possible value. The values $n = 4, 3, 2$ do not produce contradictions.

With W_1 (no rotations) and W_2 (half-turns) the pattern mesh can be defined by the contents of a single parallelogram. The pattern parallelogram of W_2 has half-turn centres at its vertices, the mid-points of its edges, and the point of intersection of its diagonals. The introduction of reflections requires rhombic or rectangular regions. In two cases, W_1^3 and W_2^4, glide reflections produce new patterns. In each pattern one of the basic parallelograms has been inserted. If rotations occur, they cover this parallelogram with copies of the smaller area indicated by dotted line(s). Reflection and glide axes are shown as full lines, they bisect these smaller areas, or the basic parallelograms in Figure 12.

For W_3 the pattern can be considered as built up of repetitions of a 60° rhombus with triple centres at its vertices. Reflection in short and long diagonals produces two variants. With W_4 a quarter-square of the basic pattern can be considered, it has half-turn centres at two vertices and quarter-turns about the other two. Reflections in the two types of

122

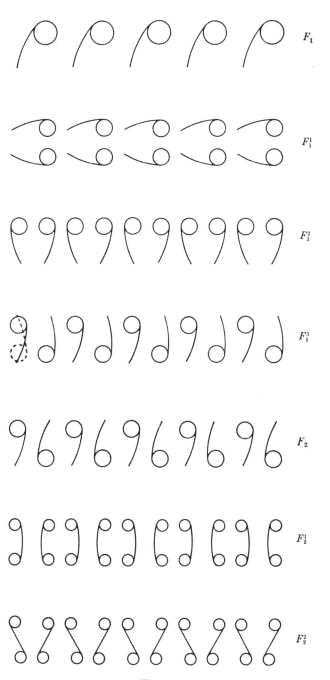

Fig. 11

123

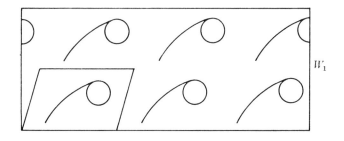

W_1'

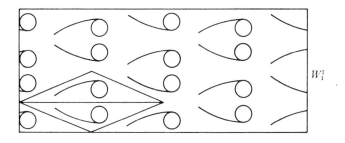

W_1^1

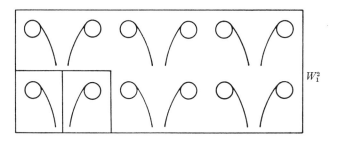

W_1^2

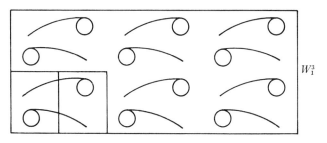

W_1^3

Fig. 12

124

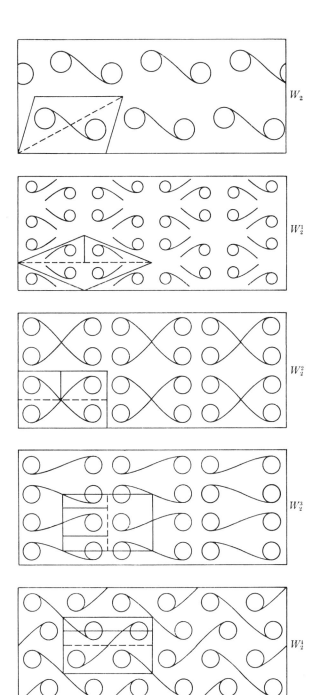

Fig. 13

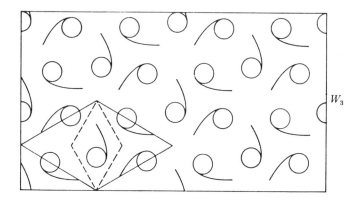

W_3

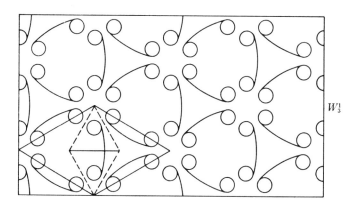

W_3^1

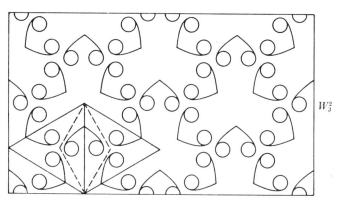

W_3^2

Fig. 14

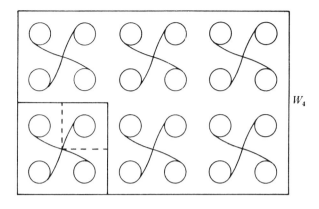

W_4

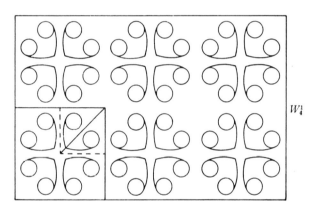

W_4^1

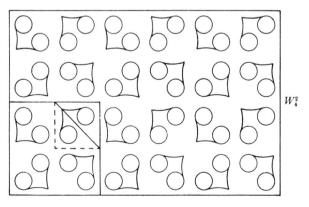

W_4^2

Fig. 15

127

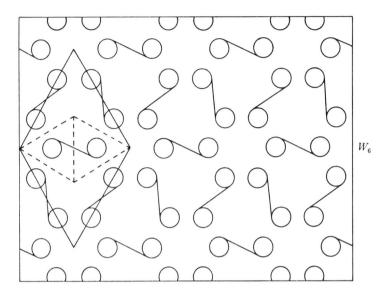

W_6

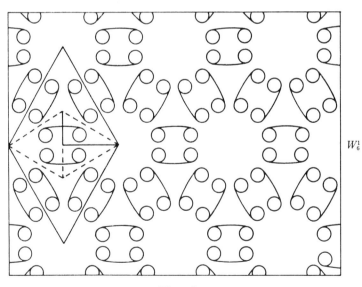

W_6^1

Fig. 16

128

diagonal lead to W_4^1 and W_4^2. The pattern W_6 can be built up by repetitions of an equilateral triangle. There are one-third-turn centres at two vertices, a one-sixth-turn centre at the third, and a half-turn centre at the mid-point of one side. Reflection in the height perpendicular to this side produces W_6^1, the richest of all the patterns in symmetry elements.

Féjes Tóth[1] gives a fairly complete account of the wallpaper patterns, while Hilbert and Cohn-Vossen[2] and Weyl[3] discuss them in outline. Weyl also derives the space analogues of the point groups found here. These are 3 in number, and are associated with the Platonic solids, being called the tetrahedral, octahedral and icosahedral groups respectively. The analogue in 3-space of the 17 wallpaper groups are the 230 space groups of crystallographic theory. It is interesting to note that the Moors used all 17 groups in ornamental decorations in the Alhambra at Granada.

References

1. L. Féjes Tóth. *Regular Figures* (Pergamon Press, 1964).
2. D. Hilbert and S. Cohn-Vossen. *Geometry and the Imagination* (Chelsea, 1952).
3. H. Weyl. *Symmetry* (Princeton, 1952).
4. I. M. Yaglom. *Geometric Transformations* (Random House, New Mathematical Library, 1962).

12

The real number system

1. *Rationals and irrationals*

We are not concerned with an axiomatic approach to the real numbers, but rather with their classification. Starting with the positive integers, the negative integers and the rationals result from the attempt to solve the equations
$$x + n = 0, \quad nx - m = 0.$$

The first defines $-n$, and the second the rational number m/n. The need for a further extension of the number system was evident to the Greeks. Geometrically the quantity $\sqrt{2}$ is capable of simple construction, but the following proof, attributed to Pythagoras, shows that there is no rational number whose square is 2.

Suppose $\sqrt{2} = p/q$, where p and q have no common factor. Then we have
$$2q^2 = p^2$$

and reason as follows. Since 2 divides the left side, it also divides the right. This is a perfect square, and so must be divisible by 4. The left is therefore also divisible by 4, and q^2 must have 2 as a factor. We have arrived at the conclusion that 2 divides both p and q, contradicting our initial assumption. The only way out of this impasse is to agree that $\sqrt{2}$ is not a rational number.

It is possible to choose rationals whose squares are as close as we please to 2. It therefore seems reasonable to introduce a new class of numbers called irrational. We can regard the irrational number $\sqrt{2}$ as the limit of a sequence of rationals, chosen so that their squares are closer and closer to 2. We looked at a continued fraction for $\sqrt{3}$ in Chapter 2. That for $\sqrt{2}$ is
$$1 + \cfrac{1}{2 + \cfrac{1}{2 + \cfrac{1}{2 + \dots}}}.$$

By using more and more terms of this fraction we get the sequence of rationals
$$\frac{1}{1}, \quad \frac{3}{2}, \quad \frac{7}{5}, \quad \frac{17}{12}, \quad \frac{41}{29}, \quad \dots,$$

130

whose squares are

$$2-1, \quad 2+\tfrac{1}{4}, \quad 2-\tfrac{1}{25}, \quad 2+\tfrac{1}{144}, \quad 2-\tfrac{1}{841}, \quad \dots .$$

The formation of successive values of the numerators is illustrated by the results

$$7 = 2 \times 3 + 1, \quad 17 = 2 \times 7 + 3, \quad \dots,$$

and so on. The same recurrence relation is used for denominators.

Other limiting processes, such as the summation of more and more terms of the series for e given below, lead to a sequence of rationals which converge to an irrational limiting value. The Wallis infinite product for π given in Chapter 14 is another example. We shall prove that π is irrational in Chapter 14; we now show that e is not a rational number.

We define e by the infinite series

$$e = 1 + \frac{1}{1!} + \frac{1}{2!} + \frac{1}{3!} + \dots,$$

and assume it is the rational number p/q. We have the result

$$q!\left(e - 1 - \frac{1}{1!} - \frac{1}{2!} - \dots - \frac{1}{q!}\right) = \frac{1}{q+1} + \frac{1}{(q+1)(q+2)} + \dots . \qquad (1)$$

Now the expression on the left is an integer, since $e = p/q$, while the right is positive and less than.

$$\frac{1}{q+1} + \frac{1}{(q+1)^2} + \dots = \frac{1}{q} < 1.$$

The contradiction causes us to dismiss the assumption of rationality.

In general it is difficult to carry through proofs of this kind. Thus we do not know if Euler's number

$$\gamma = \lim_{n \to \infty} \left\{1 + \frac{1}{2} + \dots + \frac{1}{n} - \log n\right\}$$

or $\pi^{\sqrt{2}}$ is rational or irrational.

We end this section by outlining an approach to irrational or incommensurable ratios in similiarity. The Greek geometers saw the need for a logical treatment, and their genius evolved one of astonishing modernity in outlook. We first prove that between any two real numbers α and β we can find a rational m/n. The axiom of Archimedes asserts that, given two real numbers x and y, there is an integer n such that $nx > y$. This means that we can define m, n by

$$n(\beta - \alpha) > 1, \quad m < n\beta \leqslant m + 1,$$

for if the n chosen to satisfy the first makes $n\beta$ an integer the sign $\leqslant$ still allows the second to be satisfied. The inequalities

$$\alpha < \beta - \frac{1}{n} \leqslant \frac{m+1}{n} - \frac{1}{n} = \frac{m}{n}, \quad \frac{m}{n} < \beta,$$

follow at once. This property of the rational numbers is summarised by saying that they are everywhere dense in the real number system. We now apply this result to prove the fundamental theorem concerning intercepts on a transversal by a set of parallel lines.

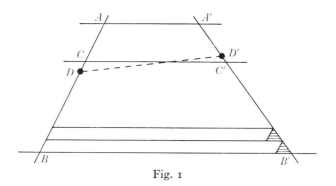

Fig. 1

Our problem is to prove that $AC/CB = A'C'/C'B'$, given that AA', BB' and CC' are parallel. If the first ratio is the rational p/q, we divide AB into $(p+q)$ equal parts, and AC will occupy just p of these. Draw lines through the points of division parallel to AA'. It is easy to prove by way of a set of congruent triangles that they cut off equal intervals on $A'B'$. Two such triangles are shaded in Figure 1. There will be p equal intervals in $A'C'$, and $B'C'$ occupies q more; the proof follows at once.

In the case where AC/AB is irrational, let us assume that

$$AC/AB < A'C'/A'B'.$$

We can find a rational r so that

$$\frac{AC}{AB} < r < \frac{A'C'}{A'B'}. \tag{2}$$

Let $AD = rAB$, and draw DD' parallel to AA'. By the result proved for the rational case $A'D' = rA'B'$. Moreover, (2) tells us that

$$AD > AC, \quad A'D' < A'C'.$$

132

This implies that DD' cuts CC', in direct contradiction to their parallelism. In a similar way we derive a contradiction from the assumption that $AC/AB > A'C'/A'B'$, and conclude that these ratios must be equal.

2. *The approximation of irrationals by rationals*

To convert p/q to decimal form we use the division process. There are q possible remainders, so after at most q steps one must recur, and the process is then cyclic. Thus rationals always give terminating or recurring decimals. The representation is unique except for the possibility exemplified by

$$0.3999\ldots = 0.4000\ldots.$$

The converse is also true; a recurring decimal necessarily represents a rational number. For example

$$0.\dot{1}2\dot{3} = \frac{123}{1000} + \frac{123}{1000^2} + \ldots = \frac{123}{1000} \cdot \frac{1}{1 - \frac{1}{1000}} = \frac{123}{999}.$$

While a decimal representation is most convenient for practical purposes, the well-known ratio $22/7$ shows that rational approximations have their uses. They are of considerable importance from a number-theoretic standpoint.

Dirichlet's theorem states that an irrational number ξ has an infinite number of rational approximations m/n, in lowest terms, and such that

$$\left| \xi - \frac{m}{n} \right| < \frac{1}{n^2}. \tag{3}$$

This is not possible for a rational number p/q in lowest terms, as

$$\left| \frac{p}{q} - \frac{m}{n} \right| = \left| \frac{pn - qm}{qn} \right| \geqslant \frac{1}{qn},$$

since the top line of the second expression is a non-zero integer unless $q = n$ and $p = m$. For property (3) to hold, n must be less than q and there are only a finite set of possibilities. We say that irrationals can be approximated to order 2 but rationals only to order 1. To prove Dirichlet's theorem we first show that, for a given integer k, there is an approximation m/n with $n \leqslant k$ such that

$$\left| \xi - \frac{m}{n} \right| < \frac{1}{nk}. \tag{4}$$

Consider the set of $(k+1)$ numbers

$$0, \quad \xi-[\xi], \quad 2\xi-[2\xi], \quad ..., \quad k\xi-[k\xi], \tag{5}$$

all less than 1, together with the set of k regions

$$(0, 1/k), \quad (1/k, 2/k), \quad ..., \quad (1-1/k, 1). \tag{6}$$

We invoke Dirichlet's pigeon-hole principle, used in Chapter 3. As each of (5) lies in one of the regions (6), there must be one region with at least 2 numbers of the set in it. Let them correspond to integers r and s, so that we have

$$(r-s)\xi-[r\xi]+[s\xi] < 1/k. \tag{7}$$

Putting $n = (r-s)$, $m = [r\xi]-[s\xi]$ and dividing (7) by n, we get (4). Should m and n have a common factor, so that $m = fM$ and $n = fN$, with M and N co-prime, (4) becomes

$$\left|\xi-\frac{M}{N}\right| < \frac{1}{fNk} < \frac{1}{Nk},$$

and so is unaltered in form.

To prove Dirichlet's theorem we suppose that there are only a finite set of approximations satisfying (3). Each one gives rise to a non-zero error, for the number ξ is not rational. Let e be the error of smallest modulus and choose $k > 1/|e|$. By the result just proved we can find an approximation satisfying (3), such that

$$\left|\xi-\frac{m}{n}\right| < \frac{1}{nk} < \frac{1}{k} < |e|.$$

This approximation, having an error less than $|e|$, cannot be one of the original set. Thus the assumption of a finite number of approximations satisfying (3) leads to a contradiction.

The most powerful approach to questions of this kind is by way of continued fractions. They can be used to show that (3) can be replaced by the stronger result

$$\left|\xi-\frac{m}{n}\right| < \frac{1}{\sqrt{5}\,n^2}.$$

It can be proved that, for some irrationals, $\sqrt{5}$ cannot be replaced by a larger number, so this result is the best possible. Hardy and Wright[2], Drobot[6] and Niven[4] deal with the continued fraction approach, while Niven[5] gives a detailed account of Dirichlet's theorem.

134

3. Algebraic numbers

We can regard the irrational $\sqrt{2}$ as defined by

$$x^2 - 2 = 0.$$

More generally we say that the algebraic equation

$$a_0 + a_1 x + a_2 x^2 + \ldots + a_m x^m = 0 \tag{8}$$

defines a set of algebraic numbers, one for each real root. The coefficients are integers or rational numbers; the numbers so defined are said to be of order m. The rationals are algebraic numbers of order unity.

Suppose we add together algebraic numbers α and β of orders m and n respectively. Then we shall prove that the result is another such number, in general of order mn. We have

$$a_m \alpha^m = -a_0 - a_1 \alpha - \ldots - a_{m-1} \alpha^{m-1}, \tag{9}$$

$$a_m \alpha^{m+1} = -a_0 \alpha - a_1 \alpha^2 - \ldots - a_{m-1} \alpha^m, \tag{10}$$

$$a_m \alpha^{m+2} = -a_0 \alpha^2 - a_1 \alpha^3 - \ldots - a_{m-1} \alpha^{m+1}. \tag{11}$$

The equation (9) can be used to eliminate α^m from (10), while (9) and (10) will remove α^m and α^{m+1} from (11). Thus we can express α^m and higher powers of α as sums of rational multiples of $1, \alpha, \alpha^2, \ldots, \alpha^{m-1}$. A similar result holds for β, and we now turn to consider powers of $(\alpha + \beta)$.

We can find rationals c_{pqr} so that

$$(\alpha + \beta)^j = c_{j,0,0} + c_{j,1,0} \alpha + c_{j,1,1} \alpha \beta + \ldots + c_{j,m-1,n-1} \alpha^{m-1} \beta^{n-1}, \tag{12}$$

for $j = 0, 1, 2, \ldots, mn$. They are obtained by expanding by the binominal theorem, and eliminating powers of α greater than $(m-1)$ and β greater than $(n-1)$, by the method outlined above. Thus the powers of $(\alpha + \beta)$ have been expressed in terms of $\alpha^s \beta^t$ for s from 0 to $(m-1)$, and t from 0 to $(n-1)$. These mn quantities can be eliminated from the $(mn+1)$ equations (12) to give the determinant

$$\begin{vmatrix} (\alpha+\beta)^{mn} & c_{mn,0,0} & \cdots & c_{mn,m-1,n-1} \\ (\alpha+\beta)^{mn-1} & c_{mn-1,0,0} & \cdots & \cdots \\ \vdots & & & \\ 1 & c_{0,0,0} & \cdots & \cdots \end{vmatrix} = 0.$$

Expansion in terms of the first column yields an algebraic equation in $(\alpha + \beta)$ of degree mn with rational coefficients.

Thus the sum of two algebraic numbers is an algebraic number, and the same holds for the other operations of arithmetic. The algebraic number system is closed, and its members form a field. There is a still more general form of closure. We might ask if the numbers x defined by the quadratic

$$x^2 + xz - y = 0, \tag{13}$$

with its coefficients positive algebraic numbers specified by

$$y^2 - 2 = 0, \quad z^2 - z - 1 = 0,$$

are still algebraic numbers. The answer can be found by eliminating y and z to get

$$x^8 + 2x^7 - x^6 - 2x^5 - 3x^4 - 4x^3 - 6x^2 + 4 = 0.$$

Two of the roots of this equation are the numbers defined by (13). This result is general, and nothing new emerges when we allow the coefficients of (8) to be algebraic numbers themselves.

However, there is another category of real numbers to which π and e belong. This class is called transcendental, and its members do not satisfy any algebraic equation like (8). We shall prove the existence of such numbers in two quite different ways. The problem of proving that a particular number is transcendental is a difficult one. Hermite proved e transcendental in 1873; Lindemann extended his method, and used

$$e^{i\pi} + 1 = 0$$

to prove π transcendental in 1882. The one general result known was discovered independently by Gelfond and Schneider in 1934, after Hilbert had propounded the problem in 1900. If α and β are algebraic, neither 0 or 1, and β is irrational, then α^β is transcendental. For instance, we can be sure that $2^{\sqrt 2}$ is transcendental, but 2^e is still in doubt, as e, although irrational, is not algebraic.

An interesting result, due to Mahler, concerns polynomials $f(x)$ that take integer values when $x = 1, 2, \ldots$. As an example consider $x(x+1)/2$, taking the values $1, 3, 6, 10, \ldots$. Mahler proved that the number

$$0 \cdot f(1)\, f(2)\, f(3) \cdots,$$

is transcendental. In our example the construction leads to the number $0 \cdot 1361015 \ldots$.

Those algebraic numbers that can be constructed using ruler and compass only are called Euclidean. We see heuristically that such numbers are combinations of quadratic surds, for intersections of lines and circles with lines and circles lead to linear or quadratic equations. The Greeks considered three problems of geometrical construction that defied all attempts at solution for 2,000 years. They were the duplication of the cube, the trisection of the angle, and the squaring of the circle. The equations to be solved are

$$x^3 - 2 = 0, \quad 4x^3 - 3x + c = 0, \quad x^2 - \pi = 0.$$

It can be shown that the first two do not reduce to quadratic form, i.e. they do not define Euclidean numbers. These problems are thus insoluble. The third demands the construction of a transcendental number, again an impossible task. Courant and Robbins[1] and Klein[3] discuss these three problems in more detail; Klein[3] proves that π is transcendental.

4. *Liouville's theorem*

In 1851 Liouville found the first demonstrably transcendental number. He proved the following theorem: an algebraic number defined by an equation of degree r is not approximable to an order higher than r. The number ξ satisfies

$$f(\xi) = a_0 + a_1\xi + \ldots + a_r\xi^r = 0.$$

We can find M so that

$$|f'(x)| < M \quad \text{if} \quad |x - \xi| < 1.$$

Consider the difference

$$f\left(\frac{p}{q}\right) - f(\xi) = f\left(\frac{p}{q}\right) = \left(\frac{p}{q} - \xi\right)f'(x), \tag{14}$$

for some value x between ξ and p/q by the first Mean Value theorem. Now

$$\left|f\left(\frac{p}{q}\right)\right| = \left|a_0 + a_1\frac{p}{q} + \ldots + a_r\frac{p^r}{q^r}\right| = \frac{\text{Integer}}{q^r} \geqslant \frac{1}{q^r},$$

since the integer cannot vanish or ξ would be rational. From this result and (14) it follows that

$$\left|\frac{p}{q} - \xi\right| > \frac{1}{Mq^r}. \tag{15}$$

If ξ is approximable to order $(r+1)$, then, for some constant c, and for infinitely many fractions p/q in lowest terms

$$\left|\frac{p}{q} - \xi\right| < \frac{c}{q^{r+1}}. \tag{16}$$

137

This requires $q < Mc$ in view of (15), so that (16) has only a finite number of possibilities.

Liouville found a class of numbers that could be approximated to an arbitrarily chosen order, and could not therefore be algebraic by the above theorem. The simplest member of this class is the sum of the series

$$\frac{1}{10^{1!}} + \frac{1}{10^{2!}} + \frac{1}{10^{3!}} + \dots$$

Suppose we wish to approximate to order k. Choose any j greater than k, and sum the first j terms. The resulting rational p/q will have $q = 10^{j!}$. The error will be

$$\frac{1}{10^{(j+1)!}} + \frac{1}{10^{(j+2)!}} + \dots = \frac{1}{q^{(j+1)}} + \frac{1}{q^{(j+1)(j+2)}} + \dots < \frac{2}{q^{(j+1)}} < \frac{2}{q^{k}},$$

as we chose j in excess of k. There are an infinity of possible values of j, so the number is approximable to order k for any k whatever.

As recently as 1955 Roth proved the much more powerful result that no algebraic number can be approximated to an order higher than 2. We have shown that all irrationals can be approximated to this order; the rationals are only approximable to order 1.

5. *Cantor's construction of transcendental numbers*

Cantor called a set of objects countable if they can be placed in one-to-one correspondence with the positive integers. He proved that the rational numbers are countable as follows. We associated with p/q the index $(p+q)$ and arrange the rationals in groups with the same index. For instance, with index 4 we have the rationals $3/1$, $2/2$, $1/3$. The numbering or counting of the rationals commences with the single member $1/1$ of index 2. The second and third are $2/1$, $1/2$ of index 3, then integers 4, 5 are allotted to the class of index 4, and so on. Fractions like $2/2$ not in their lowest terms are omitted.

A similar procedure enables us to count the system of algebraic numbers. We define as the index of the algebraic number or numbers specified by (8) the quantity

$$m + |a_0| + |a_1| + \dots + |a_m|.$$

There is no loss of generality in assuming that the a's are integers. For a given value of this index, which is necessarily an integer, there are a finite number of possible equations; these define a finite number of algebraic numbers. Some will have been defined by equations belonging to smaller indices. These are rejected, and the remainder arranged in increasing numerical order. Then, starting at the lowest index and working upwards, the process used for rationals can be repeated.

Cantor then assumed that all the real numbers in the range (0, 1) can be counted. On this assumption they can be listed as follows

$$x_1 = 0{\cdot}d_{11}d_{12}d_{13}...$$
$$x_2 = 0{\cdot}d_{21}d_{22}d_{23}...$$
$$x_3 = 0{\cdot}d_{31}d_{32}d_{33}...$$
$$x_4 = 0{\cdot}d_{41}...$$
$$.....................$$

From this set we construct a new number, not in the set itself. It commences $0{\cdot}$ and its first decimal digit is any one differing from d_{11}. Its second can be any digit differing from d_{22} and so on. This new number differs in at least one digit from each of the x's. Thus we have constructed a new number, so that our assumption that all numbers could be counted must be abandoned. Since the algebraic numbers are countable, other types (i.e. transcendental numbers) must exist.

An approach based on measure theory makes Cantor's last step unnecessary. The measure of a class of numbers is defined as the limit of the sum of a set of intervals including every number of the class. We prove that the measure of a countable class of numbers is zero. Enclose a_n, the nth member of the class, in the interval $(a_n - \epsilon/2^n, a_n + \epsilon/2^n)$. The total length of the intervals used will be

$$\epsilon + \tfrac{1}{2}\epsilon + (\tfrac{1}{2})^2\epsilon + ... = 2\epsilon,$$

a quantity that can be made as small as we choose.

Now heuristically we see that the measure of the class of all the numbers in (0, 1) must exceed zero. As the measure of the algebraic numbers in this interval, or indeed in any interval, is zero, the existence of other numbers follows. We say that almost all numbers are transcendental, since the exceptions are of measure zero.

6. *Normal numbers*

Borel introduced quite a different way of looking at the real number system. He defined as simply normal a number whose decimal expansion contains each of the digits from 0 to 9 in just $\frac{1}{10}$ of all possible positions.

Most rational numbers do not have this property but the number $0\cdot\overline{1234567890}$ is simply normal. However, like all other rationals, it fails to pass his criterion for (full) normality. This is that any 2-digit group, for example, 37 shall occur in just $\frac{1}{100}$ of all possible 2-digit selections. A 3-digit group occurs in $\frac{1}{1000}$ of all such groups and so on. We note that similar definitions could be made for a radix other than 10.

Borel proved that almost all numbers are normal to any radix, i.e. that the measure of those numbers not normal to radix r is zero. In spite of this it is not easy to prove that a given number is normal, thus π and e have defied investigation so far. Nor is it easy to write down a demonstrably normal number. Such a one is

$$0\cdot12345678910111213\ldots, \tag{17}$$

where the decimal digits arise by placing the ascending integers in juxtaposition.

In order to prove this we concentrate our attention on the triplet 327, and count its occurrences in the integers from $10^{(k-1)}$ to $10^k - 1$. These occurrences are of four types:

(i) $|327 \leftarrow (k-3) \text{ digits} \rightarrow |$,

(ii) $| \leftarrow l \text{ digits} \rightarrow 327 \leftarrow (k-l-3) \text{ digits} \rightarrow |$,

(iii) $| \leftarrow (k-1) \text{ digits} \rightarrow 3||27 \leftarrow (k-2) \text{ digits} \rightarrow |$,

(iv) $| \leftarrow (k-2) \text{ digits} \rightarrow 32||7 \leftarrow (k-1) \text{ digits} \rightarrow |$,

and type (ii) occurs for values of l from 1 to $(k-3)$. In case (i) there are 10 possible digits in each of the $(k-3)$ arbitrary positions, so we have $10^{(k-3)}$ triplets of this type. In case (ii) the first digit must not be zero, but takes all other values; allowing for the various values of l we get $9.(k-3).10^{(k-4)}$ possibilities. In case (iii) a fraction of $\frac{1}{10}$ of the numbers preceding a number starting with 27 will end in 3. Hence we get $\frac{1}{10}$ times $10^{(k-2)}$ or $10^{(k-3)}$ of this type. Similarly case (iv) gives $\frac{1}{100}$ times $10^{(k-1)}$ triplets 327. The grand total is found to be $(9k+3).10^{(k-4)}$. We must now count the possible number of triplets. Each digit except for the last two is the first one of a triplet. The number of digits is k times the number of integers, this is $k(10^k - 10^{(k-1)})$ or $9k.10^{(k-1)}$. Thus the proportion of 327's among the triplets will be

$$\frac{9k+3}{9000k - 2.10^{-k+4}},$$

a quantity that tends to $\frac{1}{1000}$ for large k. Thus, over all values of k, 327 occurs in the correct proportion of cases. A similar proof applies for

140

quadruples and so on. A few triples like 991 will occur at the junctions of different 'k' blocks, but these instances are such a small part of the whole as to be negligible. Niven[4] gives an excellent account of normality. We note that the number (17) is irrational, for its digits cannot recur. Mahler's result, quoted in Section 3, states that it is transcendental. So this number has two of the properties possessed by almost all real numbers; it is both transcendental and normal.

References

1. R. Courant and H. Robbins. *What is Mathematics?* (Oxford, 1943).
2. G. H. Hardy and E. M. Wright. *An Introduction to the Theory of Numbers*, 4th edition (Oxford, 1960).
3. F. Klein. *Famous Problems of Elementary Geometry* (Dover, 1956).
4. I. Niven. *Irrational Numbers*, Carus Monograph no. 11 (Wiley, 1956).
5. I. Niven. *Numbers: Rational and Irrational* (Random House New Mathematical Library, 1961).
6. S. Drobot. *Real Numbers* (Prentice-Hall, 1964).

13

A theorem of combinatorial geometry

1. *Helly's theorem in two dimensions*

Helly's theorem belongs to the somewhat ill-defined sphere of combinatorial geometry. It also has applications in analysis, particularly in the theory of approximations. We shall prove the result in a simple form, and mention an extension. The n-dimensional version is stated later.

We have a finite collection of convex regions such that any 3 of the regions have a point in common. The theorem states that they all have a point in common. Figure 1(a) shows that the number 3 cannot be replaced by 2. Each pair of the circles has a common point, but no point

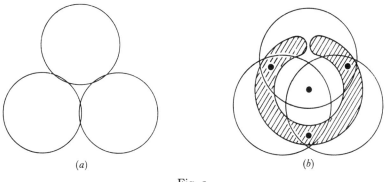

(a) (b)

Fig. 1

belongs to all 3. Figure 1(b) shows that convexity of the regions is essential. Each pair of circles have a point in common with the shaded zone, and there is a point common to all 3 circles. However, there is no point lying in all 4 regions.

We start by proving the result for 4 regions R_1 to R_4. Let A_1 lie in R_2, R_3 and R_4, A_2 lie in R_1, R_3 and R_4, and so on. Then every part of the triangle A_1, A_2, A_3 lies in R_4, for all its vertices are in this region, and the region is convex. Similar results hold for the other 3 triangles. There are two essentially different dispositions of the points A_1 to A_4. Either they form a convex quadrilateral, or one of them lies inside the triangle formed by the other three. In the first case the point of intersection of the diagonals belongs to all 4 triangles, and hence lies in all the regions. In the second case, if A_4 lies inside $A_1 A_2 A_3$ then A_4 belongs to all 4 triangles and is again within R_1 to R_4.

We now assume that the theorem holds for all sets of n regions. Consider $(n+1)$ regions R_1 to R_{n+1} satisfying the conditions of the theorem. Since R_n, R_{n+1} and R_1 have a common point, the intersection R of the first two is not empty. Moreover, R, whose points lie in both convex regions, is itself convex; this is proved as follows. A necessary and sufficient condition for convexity is that the joint of any 2 points in the region lies entirely within it. If we select points 1 and 2 of R they lie in convex region R_n, and therefore all points of line (1, 2) lie in R_n. Similarly line (1, 2) lies entirely in R_{n+1} and hence also in R.

Consider the set of n convex regions R_1 to R_{n-1} and R. By the result just proved for 4 regions we see that R_i, R_j, R_n and R_{n+1} have a common point, for any 3 of them have such a point. Hence too R_i, R_j and R have a point in common because of the way R is defined. This is true for all i, j from 1 to $(n-1)$, therefore this set of n regions satisfies the conditions of the theorem. The theorem is assumed to be true for any set of n regions, and they thus have a point in common. This lies in R, and hence belongs to both R_n and R_{n+1}, i.e. all $(n+1)$ regions have a common point, and our inductive proof is complete.

The theorem is also true for an infinite set of regions provided they are bounded. This assumption was not made above, and our proof applies to any finite set of regions, bounded or unbounded.

2. *A centring theorem*

We consider chords PQ of a closed convex curve through a fixed point O. In the special case where O is the mid-point of each chord we have a centrally symmetric curve. We shall prove that, in all cases, O can be chosen so that the centre of every chord is within one sixth of its length of O. Put otherwise, we can always find O so that both PO and OQ are greater than or equal to $\frac{1}{3}PQ$.

In Figure 2(a) the point A is joined to each point X of the curve and $AY = \frac{2}{3}AX$, so that Y generates a similar and similarly situated convex curve. In Figure 2(b) we have applied this construction at any 3 points A, B and C. We next show that G, the centroid of triangle ABC lies inside or on all 3 derived convex curves. The mid-point L of BC lies inside or on the outer curve, since it is convex. It is a fact of elementary geometry that $AG = \frac{2}{3}AL$. Hence AG is less than or equal to AY, i.e. G lies inside or on the derived curve at A. By symmetry, it does so for the curves at B and C.

Now we apply Helly's theorem to the infinite set of bounded convex curves obtained from all points of the original convex curve by the '$\frac{2}{3}$' construction. Any 3 have a point in common by the proof just given.

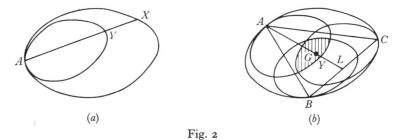

(a) $\qquad\qquad$ (b)

Fig. 2

Hence all have a point O in common. Draw a chord PQ through O, and consider the '$\frac{2}{3}$' curve for P. It contains O, and therefore OQ equals or exceeds $\frac{1}{3}PQ$. In exactly the same way we show that PO does so too.

For some convex regions the fraction $\frac{1}{3}$ can be increased, thus $\frac{1}{2}$ is possible for the ellipse. For the equilateral triangle no such improvement in choosing a 'centre' is possible. It has 3 chords trisected at its in-centre, any other point has chords through it divided in a ratio less than $\frac{1}{3}$.

3. *A theorem of Blaschke*

We first define the support line at any point P of a closed convex curve. It is any line through P with all points of the convex region lying on one side of it. At points where the slope is continuous it is the tangent at P. If there is a slope change at P, any of a pencil of lines lying between the two tangents will satisfy the condition.

The breadth of the curve in a given direction is the distance between the unique pair of parallel support lines perpendicular to this direction. Figure 3 illustrates three cases of this definition of breadth. As the chosen direction varies the breadth will change. We call the minimum value taken the width of the curve. Thus a curve of width unity can just be covered by a parallel-sided strip one unit wide.

Blaschke proved that a convex curve of width 1 or more always contains a circle of radius $\frac{1}{3}$. That this figure cannot be increased can be seen by considering the equilateral triangle. If its height is unity, the largest possible circle contained in it is its in-circle of radius $\frac{1}{3}$.

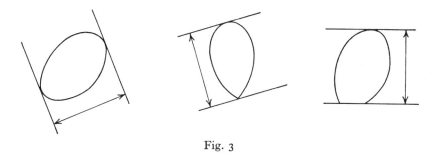

Fig. 3

We shall show that the point O of the last section is at least $\frac{1}{3}$ of a unit from every point A of the curve. It then follows that we have a circle centre O with the Blaschke property.
Taking a support line l' in Figure 4 parallel to the support line l at A, let it meet the curve at Q. Draw the chord QOP, cutting l at P'. The line XOY is perpendicular to l and l'.

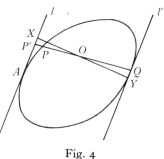

Fig. 4

We have $OP/PQ \geqslant \frac{1}{3}$, and adding equal quantities to top and bottom lines, we see that $OP'/P'Q \geqslant \frac{1}{3}$. Then by similar triangles, $OX/XY \geqslant \frac{1}{3}$, and, as XY is at least 1, this gives $OX \geqslant \frac{1}{3}$. Finally, AO equals or exceeds OX, and the theorem is proved.

4. *Krasnosellskiĭ's theorem*

We are concerned with non-convex polygons with the property of being star-shaped with regard to an interior point P. This means that we can see the whole of each side from P; or that, if we join P to any point Q of the boundary in Figure 5 (a), the half-line PQ does not cut the polygon again.

The second diagram illustrates a region possessing no point with respect to which it is star-shaped. The triangles formed by the dotted extensions of the sides would each have to contain any point from which all sides can be seen, and they do not intersect.

Krasnosellskii showed that, if each set of three sides is visible from some point, then all the sides can be seen from a suitably chosen point.

In Figure 6(a) we allot a clockwise direction of rotation to the sides, so that the interior lies to the right of each side. To each side corresponds the half-plane lying on its right; one has been shaded in the diagram. These half-planes are unbounded convex regions. A point

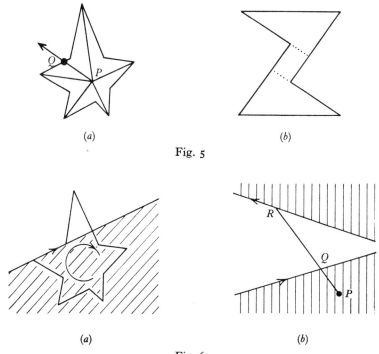

(a)   (b)

Fig. 5

(a)   (b)

Fig. 6

from which a side can be seen must lie in the half-plane corresponding to that side. Hence all sets of three half-planes have a point in common. It follows that they all have a common point P. It is easy to prove that this lies inside the polygon. We show that a ray PQ cannot have another intersection besides Q with the boundary, i.e. that all sides can be seen from P.

The proof is by *reductio ad absurdum*; we assume that PQ cuts the boundary again at R. In Figure 6 (b) there is no point of the boundary between Q and R. Hence QR is outside the polygon, and the half-plane defined by the side containing R lies on the opposite side of R to Q. But this means that it cannot contain P, a contradiction. Suppose that there are other points of the boundary between Q and R; let S be the one nearest Q. In exactly the same way we see that P lies outside the half-plane with S on its boundary.

The general non-convex polygon may have no point like P. If it has two such points, called P and Q, it is easy to see that a side visible from both P and Q remains visible from all points of the interval PQ. In fact the set of points with regard to which a region is star-shaped is either empty or convex.

5. *Jung's theorem and covering problems*

We call the maximum breadth of a convex figure its diameter. In fact we can define diameter for a quite general set of points as the maximum distance separating any pair of these points. Jung's theorem states that a plane point set of diameter unity can be covered by a circle of radius $1/\sqrt{3}$.

We consider a set of 3 points of diameter 1 or less. If they form an obtuse triangle they are included in a semi-circle on the longest side as diameter, its radius is not greater than $\frac{1}{2}$ and is therefore less than $1/\sqrt{3}$. Consequently the 3 circles of radius $1/\sqrt{3}$ with centres at the vertices all contain the mid-point of the longest side. If the triangle is acute it has one angle A lying between $60°$ and $120°$, so that its sine equals or exceeds $\frac{1}{2}\sqrt{3}$. The radius of its circumcircle is given by

$$2R = \frac{a}{\sin A}$$

and $a \leqslant 1$. Thus $R \leqslant 1/\sqrt{3}$, and this time the 3 circles introduced above all contain the circumcentre.

Returning to our set of points we draw a circle of radius $1/\sqrt{3}$ centred on each; by the results just proved and Helly's theorem they possess a common point. This point is no further than $1/\sqrt{3}$ from any of the set, so it is the centre of a covering circle of this radius. We call this circle a universal cover for sets of diameter unity. An equilateral triangle of side unity has a circumradius of $1/\sqrt{3}$, so this is the smallest possible such circle. There are other smaller universal covers, we shall prove that a regular hexagon inscribed in this circle is one. The distance between a pair of parallel sides of this hexagon is unity. The problem of the universal cover of smallest area is not yet solved. The second diagram of Figure 7, a regular hexagon with two corners removed, is another cover, and still smaller ones have been found.

To prove the result for the regular hexagon we start with three pairs of distant parallel lines with the set of points inside the large hexagon they define. The lines are chosen so that angles of this hexagon are all $\frac{2}{3}\pi$. Sides are moved in parallel to themselves until they encounter a point of the set. We arrive at an equiangular hexagon whose parallel sides are not more than unit distance apart.

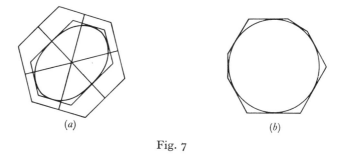

(a)   (b)

Fig. 7

There was one degree of freedom in the choice of the original lines, we define the angle that one of the three directions makes with the horizontal as θ. Lengths of sides of the hexagon will be continuous functions of θ. We consider a pair of opposite sides $a(\theta)$ and $b(\theta)$ and define the continuous function

$$d(\theta) = a(\theta) - b(\theta).$$

For a given value θ_0 we assume that a exceeds b, so that $d(\theta_0) > 0$. We have

$$d(\theta_0 + \pi) = a(\theta_0 + \pi) - b(\theta_0 + \pi) = b(\theta_0) - a(\theta_0) = -d(\theta_0),$$

since an increase of θ by π interchanges opposite sides. Thus $d(\theta)$ is of opposite signs at θ_0 and at $(\theta_0 + \pi)$. By a well-known result for continuous functions it must take the value zero at some intermediate point. So we can find an equiangular hexagon with an opposite pair of sides equal. It is a matter of simple geometry to prove that all three pairs of opposite sides are necessarily equal, i.e. that it has central symmetry.

Through the centre we draw the three perpendiculars to pairs of opposite sides. These are perpendicular to the sides of a regular hexagon with the same centre and of circumradius $1/\sqrt{3}$ illustrated in Figure 7(a). The centre is at a distance of $\frac{1}{2}$ from the sides of the regular hexagon.

148

Its distance from a side of the centrally symmetric hexagon circumscribing the set cannot exceed $\frac{1}{2}$. Otherwise this hexagon would possess a pair of parallel sides more than unit distance apart. So the regular hexagon covers the circumscribing hexagon, and hence the set of diameter unity.

6. *Some n-dimensional theorems*

We give without proofs extensions of some of the results discussed above. Helly's theorem states that, if a set of bounded convex regions in n-space is such that any $(n+1)$ have a point in common, then a point can be found lying in all of them.

Blaschke's theorem for a convex body of width exceeding 1 in n-space states that it contains a sphere of diameter

$$\frac{\sqrt{(n+2)}}{n+1} \ (n \text{ even}) \quad \text{or} \quad \frac{1}{\sqrt{n}} \ (n \text{ odd}).$$

Jung's theorem states that a hypersphere of diameter

$$\sqrt{\left(\frac{2n}{n+1}\right)},$$

is a universal cover for a set of points of unit diameter. The regular n-simplex of side

$$\sqrt{\{\tfrac{1}{2}n(n+1)\}},$$

also provides a universal cover. In 2-space an equilateral triangle of side $\sqrt{3}$ will cover the regular hexagon found in Section 5, so we have a proof of the simplex result in the plane.

7. *Approximating a function by polynomials*

We have n points (x_i, y_i) lying on the curve $y = f(x)$, and seek the 'best' approximation to these points by a straight line. Any line $y = mx + c$ will give an error

$$e_i = |y_i - mx_i - c|,$$

at the ith point. We choose as best the line that minimises the maximum error, i.e. the maximum value of the set e_1 to e_n.

149

For three points the optimum solution is shown in Figure 8(a). The absolute values of errors are equal and they alternate in sign. Assuming that x_1, x_2, x_3 are in ascending order of magnitude we solve the three equations

$$y_1 = mx_1 + c + \epsilon,$$
$$y_2 = mx_2 + c - \epsilon,$$
$$y_3 = mx_3 + c + \epsilon,$$

for m, c and ϵ. It is not possible to decrease the error at point 1 without increasing the error at either 2 or 3. A decrease at 1 means that the line moves up there; if it stays fixed or moves down at 2 it moves down at 3 so

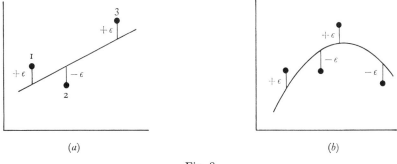

(a)   (b)

Fig. 8

increasing the error there. Otherwise the line must go up at 2, with an increase in 2's error. Attempts to decrease the maximum absolute error of the set are therefore unavailing.

To determine the optimum line for n points we first carry out this process for all possible sets of 3. Each one gives an ϵ, and one set will have a value ϵ_0 of maximum modulus. Then this line is optimum, giving ϵ_0 as the smallest possible worst error for this set of data.

The proof runs as follows. We associate with each of the n points a line joining $(x_i, y_i - \epsilon_0)$ and $(x_i, y_i + \epsilon_0)$. A typical point and the associated vertical line are shown in Figure 9(a). The solution

$$y = mx + c$$

must cut this vertical line if no error exceeds ϵ_0. This leads to the inequalities

$$y_i - mx_i - \epsilon_0 \leqslant c \leqslant y_i - mx_i + \epsilon_0.$$

The second diagram is one in which each line of the (x, y) plane has its 'c' plotted against its 'm'. The above inequalities imply that all possible

150

lines with error less than ϵ_0 lie in the shaded region between a pair of parallel lines in the (m, c) plane.

Now there are n vertical intervals, and a corresponding set of n shaded strips in the (m, c) diagram. There is a line cutting any set of 3 vertical intervals, for each set is fitted by a line with an error not exceeding ϵ_0. So any set of 3 of these convex strips have a common point. In most cases a set of 3 will have a common area, but in the set that gave ϵ_0 there will be only a single point in all 3 regions. Helly's theorem tells us that all n regions have a common point; from the observation just

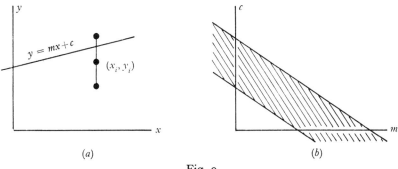

(a)   (b)

Fig. 9

made it must correspond to the unique (m, c) determined by the worst case. Since the point lies in all other regions in the (m, c) diagram, the corresponding line cuts all vertical lines in the (x, y) plane. Thus this line gives a maximum error of ϵ_0, and as one set of 3 points actually realises this maximum, there is no better fit.

In practice there may be far too many sets of 3 to make this plan possible, and iterative procedures have been worked out to avoid evaluating sets of 3 unnecessarily. Our proof above is an existence theorem, not a procedure we would necessarily employ on extensive data.

Figure 8(b) shows 4 points fitted with equal errors of alternating sign by the curve
$$y = A + Bx + Cx^2,$$
a parabola. Four points are now needed to determine A, B, C and ϵ. The (m, c) plane is replaced by an (A, B, C) 3-space. The three-dimensional form of Helly's theorem is now available to assure us that the best parabolic fit is given by trying all sets of 4 and selecting the parabola for that set giving the greatest numerical value of ϵ. Here, even more than in the previous case, some shortcutting procedure is desirable. The method

151

is quite general and indicates the best fit of degree k. This is obtained by fitting all possible sets of $(k+1)$ points with equal errors of alternating signs. The worst such set gives the curve of degree k appropriate to all data points.

Topics akin to those we have discussed in earlier sections are treated admirably by Hadwiger, Debrunner and Klee[2], Lyusternik[3] and Yaglom and Boltyanskii[4]. Eggleston's[1] book is more difficult, but it is a standard work on convexity.

References

1. H. G. Eggleston. *Convexity*. Cambridge Tract no. 47 (Cambridge, 1958).
2. H. Hadwiger, H. Debrunner and V. Klee. *Combinatorial Geometry in the Plane* (Holt, Rinehart and Winston, 1964).
3. L. A. Lyusternik. *Convex Figures and Polyhedra* (Dover, 1963).
4. I. M. Yaglom and V. G. Boltyanskii. *Convex Figures* (Holt, Rinehart and Winston, 1961).

14

The number π

1. *Historical background*

The number π has been the subject of a series of investigations spanning a period of more than 2,000 years. Its early importance lay in its relationship to the circle. With the development of the calculus its fundamental nature became apparent, and the connection with the circle is today seen to be of secondary importance. In Biblical times the value 3 was in use as an approximation; it is implied by verse 23 of I Kings 7 and verse 2 of II Chronicles 4. Since then a host of better approximations have been devised. The ratio 22/7 is the best known and, apart from decimal values, no other is in use today. Among many contenders in the past we mention $\sqrt{10}$. Although not a particularly good approximation, it was used in India, China and throughout the ancient world. We shall mention later the value 355/113, a very good approximation, discovered by Tsu Ch'ung-chih in about 470 A.D. The Indian mathematician Ramanujan, discovered by Hardy while working in obscurity in his own country, found the approximation $\sqrt[4]{(2143/22)}$, a quantity in error by 10^{-9}.

Serious attempts to compute π go back at least to Archimedes. By inscribing and circumscribing regular polygons of 96 sides in and around a circle he proved that

$$3\tfrac{10}{71} < \pi < 3\tfrac{1}{7}.$$

In this, as in so many other respects, the Greek approach is characteristically modern.

A long list of successively more accurate computations could be given, we shall mention only a few. Vieta, who found the beautiful formula discussed in Section 2, determined its value to 10 decimal places. Like most of the early computers he used a method based on the length of side of a regular n-gon inscribed in a circle. In Figure 1, AB, the side of an n-gon, is of length s_n, while BC, corresponding to $2n$ sides is s_{2n}. We have the results

$$OD^2 = 1 - \tfrac{1}{4}s_n^2, \quad s_{2n}^2 = \tfrac{1}{4}s_n^2 + (1 - OD)^2,$$

leading to the recurrence relation

$$s_{2n} = \sqrt{\{2 - \sqrt{(4 - s_n^2)}\}}.$$

Each application doubles the number of sides of the inscribed polygon. After enough steps we have N sides, and assume that $2\pi = N s_N$.

Machin, in about 1700, used formula (3) of the next section to derive π to 100 decimals. From this time formulae of similar type were used, and Shanks determined 607 decimals in 1853. The digital computer has now taken over; its value has recently been determined to no less than 100,000 decimal digits. Such results enable counts of digit patterns to be made in connection with tests of normality (see Section 6 of Chapter 12). It is not yet known if π is a normal number.

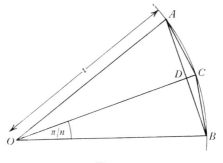

Fig. 1

We turn to the status of π in the real number system. Greek attempts to draw a square equal in area to a circle by ruler and compass methods all failed. Lambert proved that π was irrational in 1761, and by then it was suspected that the problem was insoluble. This was proved by Lindemann in 1882, when he showed that π is a transcendental number. We prove π's irrationality later; a proof of its transcendental nature is given by Klein[2].

2. *Some formulae for* π

Vieta gave one of the first instances of an infinite product in 1593; it is

$$\frac{2}{\pi} = \frac{\sqrt{2}}{2} \cdot \frac{\sqrt{(2 + \sqrt{2})}}{2} \cdot \frac{\sqrt{\{2 + \sqrt{(2 + \sqrt{2})}\}}}{2} \dots.$$

Successive factors converge to 1 with reasonable speed, the 25th being $1 - 10^{-15}$. Proof follows from the formulae

$$\sin \theta = 2 \cos \frac{\theta}{2} \sin \frac{\theta}{2}, \quad \cos \frac{\theta}{2} = \frac{\sqrt{(2 + 2 \cos \theta)}}{2}. \tag{1}$$

Repeated use of the first gives

$$1 = \sin \frac{\pi}{2} = 2^{n-1} \cos \frac{\pi}{4} \cos \frac{\pi}{8} \dots \cos \frac{\pi}{2^n} \sin \frac{\pi}{2^n}.$$

Then the second formula is used to evaluate the cosines giving

$$\frac{1}{2^{n-1} \sin \pi/2^n} = \frac{\sqrt{2}}{2} \cdot \frac{\sqrt{(2 + \sqrt{2})}}{2} \cdot \dots \cdot \frac{\sqrt{\{2 + (\sqrt{2} + \dots)\}}}{2},$$

the last term containing $(n-1)$ nested root signs. Letting n tend to infinity and replacing sine by angle gives Vieta's formula.

Our next result, another infinite product, was discovered by Wallis (1616–1703). It reads

$$\frac{2}{\pi} = \frac{1 \cdot 3}{2 \cdot 2} \cdot \frac{3 \cdot 5}{4 \cdot 4} \cdot \frac{5 \cdot 7}{6 \cdot 6} \dots \cdot \frac{(2n-1)(2n+1)}{2n \cdot 2n} \dots .$$

Convergence is very slow, the 50th pair of products giving $1 - 10^{-4}$. To prove the result we first obtain a certain reduction formula by integrating by parts. We have

$$I_n = \int_0^{\frac{1}{2}\pi} \sin^n \theta \, d\theta = [-\cos \theta \sin^{n-1} \theta]_0^{\frac{1}{2}\pi} + \int_0^{\frac{1}{2}\pi} (n-1) \sin^{n-2} \theta \cos^2 \theta \, d\theta$$

$$= (n-1)(I_{n-2} - I_n),$$

on replacing $\cos^2 \theta$ by $1 - \sin^2 \theta$.
This gives

$$I_n = \frac{n-1}{n} I_{n-2}. \tag{2}$$

Noting that $I_0 = \frac{1}{2}\pi$, $I_1 = 1$, we obtain

$$I_{2n} = \frac{2n-1}{2n} \cdot \frac{2n-3}{2n-2} \dots \cdot \frac{1}{2} \cdot \frac{\pi}{2},$$

$$I_{2n+1} = \frac{2n}{2n+1} \cdot \frac{2n-2}{2n-1} \dots \cdot \frac{2}{3} \cdot 1.$$

Since $\sin \theta < 1$ we have

$$I_{2n} > I_{2n+1} > I_{2n+2},$$

for the integrand decreases as n increases. From (2) we see that, with increase of n, the ratio I_{2n}/I_{2n+2} tends to 1. The result just proved shows that this is also true of I_{2n}/I_{2n+1}, and this ratio yields the Wallis product.

155

Brouncker's continued fraction for π is of about the same period; it is

$$\frac{4}{\pi} = \cfrac{1}{1 + \cfrac{1^2}{2 + \cfrac{3^2}{2 + \cfrac{5^2}{2 + \dots}}}}.$$

It is easily derived from the continued fraction for $\tan^{-1} x$, given in Chapter 2. However, this result was found by Euler a century later, and we do not know what method Brouncker used.

Gregory's series (1671),

$$\tan^{-1} x = x - \tfrac{1}{3}x^3 + \tfrac{1}{5}x^5 - \dots,$$

can be derived by expanding $1/(1 + x^2)$ by the Binomial Theorem, and integrating the result. Putting $x = 1$ we obtain the Leibnitz series

$$\tfrac{1}{4}\pi = 1 - \tfrac{1}{3} + \tfrac{1}{5} - \tfrac{1}{7} + \dots.$$

Convergence is so slow as to make the result useless for computational purposes. However, the addition formula for $\tan(x+y)$ enables us to prove that

$$\tfrac{1}{4}\pi = 4 \tan^{-1} \tfrac{1}{5} - \tan^{-1} \tfrac{1}{239}. \tag{3}$$

Gregory's series converges rapidly for the second term, and is practicable for the first. This formula of Machin's has been used in several attempts to find π to many decimal digits.

Euler in about 1745 considered the problem of summing the inverse squares of the integers, thus discovering a set of series formulae for π. His reasoning, discussed by Polya[4], was on the following lines. Seeking an algebraic equation with roots of the form $1/n^2$, he was led to assume

$$\sin x = Ax \left(1 - \frac{x^2}{\pi^2}\right)\left(1 - \frac{x^2}{2^2\pi^2}\right)\dots. \tag{4}$$

As $\sin(x)$ vanishes for $x = 0, \pm\pi, \pm 2\pi, \dots$ an extension of the Remainder Theorem makes this result plausible. For small x, the product is close to x, so that $A = 1$.

Taking logs of both sides, after dividing by x and replacing $\sin(x)$ by its Taylor series expansion about zero, gives

$$\log\left\{1 - \frac{x^2}{3!} + \frac{x^4}{5!} - \dots\right\} = \sum_{j=1}^{\infty} \log\left(1 - \frac{x^2}{j^2\pi^2}\right).$$

The logarithmic expansion is then applied to both sides, it being assumed that this is justifiable for the double series that results from the right-hand side. We get

$$-\frac{x^2}{6}-\frac{x^4}{180}-\dots = -\frac{x^2}{\pi^2}\sum_{j=1}^{\infty}\frac{1}{j^2}-\frac{x^4}{2\pi^4}\sum_{j=1}^{\infty}\frac{1}{j^4}-\dots,$$

and comparison of powers of x yields the series

$$\frac{\pi^2}{6} = \frac{1}{1^2}+\frac{1}{2^2}+\frac{1}{3^2}+\dots,$$

$$\frac{\pi^4}{90} = \frac{1}{1^4}+\frac{1}{2^4}+\frac{1}{3^4}+\dots.$$

The first solved Euler's problem, the second converges rapidly enough to provide a useful way of determining π. It can be improved by noting that

$$1-\frac{1}{2^4}+\frac{1}{3^4}-\frac{1}{4^4}+\dots = \frac{\pi^4}{90}\left(1-\frac{2}{16}\right) = \frac{7\pi^4}{720},$$

since the even terms of the original series add up to give $\frac{1}{16}$th of the full sum. The nth term of the new series is of the same magnitude as in the old; however, the alternation of signs is valuable. It is easy to prove that, because of the steadily decreasing terms, when the series is truncated the error incurred is of smaller magnitude than the first term omitted.

3. *Three geometrical approximations*

While π, a transcendental number, cannot be constructed by ruler and compass, attempts to square the circle have led to many geometrical approximations. The following was discovered by Kochansky, a Polish monk, in 1685. In Figure 2 the circle is of unit radius, and we construct angle BOC equal to $30°$. The line CD is 3 units long, and AD is very nearly equal to π. We have $CB = 1/\sqrt{3}$ so that

$$AD = \sqrt{\{2^2+(3-1/\sqrt{3})^2\}} = \sqrt{(\tfrac{40}{3}-\sqrt{12})} = 3\cdot141533,$$

contrasting with the true value $3\cdot141593$.

Before dealing with the second approximation we digress to obtain a continued fraction expansion for π with unit numerators. The decimal part, multiplied by 10^8 is first divided into 10^8. The remainder now becomes a divisor, and so on. This is of course the process for finding a highest common factor.

$$
\begin{array}{r}
14,159,265)100,000,000(7 \\
99,114,855 \\
885,145)14,159,265(15 \\
13,277,175 \\
882,090)885,145(1 \\
882,090 \\
\hline
3,055
\end{array}
$$

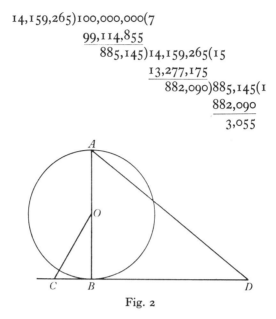

Fig. 2

The result is the continued fraction

$$3+\cfrac{1}{7+\cfrac{1}{15+\cfrac{1}{1+\dots}}}$$

Had more terms been required we would have started with more digits in π. The next term is $\frac{1}{292}$, and its small size indicates that neglecting it, together with all subsequent fractions, causes only a small error in the approximation. Truncating at successively lower levels we find

$$
\begin{aligned}
3 &= 3\cdot0, \\
\tfrac{22}{7} &= 3\cdot1429, \\
\tfrac{333}{106} &= 3\cdot141509, \\
\tfrac{355}{113} &= 3\cdot14159292.
\end{aligned}
$$

These values are to be compared with $3\cdot14159265$, and the last gives excellent agreement.

Figure 3 illustrates Gelder's construction based on the ratio $\frac{355}{113}$. The circle is of unit radius, and $OB = \frac{7}{8}$, $AC = \frac{1}{2}$, CD is perpendicular to

158

AO, while *CE* is parallel to *BD*. Then *AE* is near $(\pi-3)$. The proof is straightforward, we first note that $AB = \sqrt{(113)}/8$. By similar triangles

$$\frac{AD}{AO} = \frac{AC}{AB} \quad \text{and} \quad \frac{AE}{AD} = \frac{AC}{AB},$$

the 'product' of these expressions gives

$$\frac{AE}{AO} = \frac{AC^2}{AB^2} = \frac{16}{113} = \frac{355}{113} - 3.$$

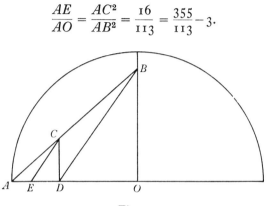

Fig. 3

The construction of Figure 4 is due to Snell, the discoverer of the law of refraction: $\sin(i) = \mu \sin(r)$. It gives an approximation to the length of circular arcs of small angle. To construct a line of approximately the

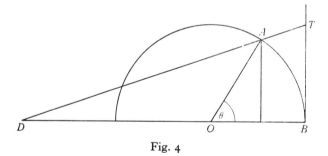

Fig. 4

same length as arc *AB*, produce the unit length *BO* to *D* so that $BD = 3$. The line *DA* cuts the tangent at *B* in *T*, and *BT* is nearly equal to θ. By similar triangles

$$\frac{BT}{3} = \frac{\sin\theta}{2+\cos\theta} = \frac{1}{3}\left(\theta - \frac{\theta^5}{180} + \dots\right).$$

The use of Taylor's series, followed by simple division, leads to the

/

159

latter expression. The neglect of all but the first term is justified for small θ. Applied to an angle $\frac{1}{16}\pi$, readily determined by repeated bisection, the construction gives a good approximation to π.

4. *The irrationality of π*

If π is rational, then π^2 is equal to p^2/q^2 and is rational too. We assume the latter result, and arrive at a contradiction.

As a preliminary we consider

$$f(x) = \frac{x^n(1-x)^n}{n!} = u(x).v(x) \quad \text{with} \quad u(x) = \frac{x^n}{n!}.$$

We are interested in the values of the derivatives of $f(x)$ at $x = 0$ and 1. Because of symmetry about $x = \frac{1}{2}$ those of even order are equal, those of odd order differ only in sign. We shall prove that all derivatives at these points are signed integers or zero. The derivatives of u are

$$u^r(0) = 0 \quad (r \neq n),$$

$$u^n(0) = 1.$$

Using the Leibnitz formula for the derivatives of a product it is easy to show that

$$f^r(0) = 0 \quad (r < n \quad \text{or} \quad r > 2n),$$

$$f^r(0) = \binom{r}{n} v^{r-n}(0) = (-1)^{r-n} \binom{r}{n} n(n-1)\ldots(2n-r+1) \quad (n < r \leqslant 2n)$$

$$= 1 \quad (r = n).$$

Thus the result has been proved.

We next consider the expression

$$I = \int_0^1 f(x) \sin \pi x \, dx,$$

this is twice integrated by parts to give

$$I = \frac{f(0)+f(1)}{\pi} - \frac{1}{\pi^2} \int_0^1 f''(x) \sin \pi x \, dx.$$

As $f^{2n+2}(x) \equiv 0$, repeated application of this result leads to

$$\pi I = \left[f(0)+f(1) - \frac{f''(0)+f''(1)}{\pi^2} + \ldots + (-1)^n \frac{f^{2n}(0)+f^{2n}(1)}{\pi^{2n}} \right].$$

160

Our assumption of rational $\pi^2 = a/b$ implies

$$\pi a^n I = a^n[f(0)+f(1)] - a^{n-1}b[f''(0)+f''(1)] + \ldots$$
$$+ (-1)^n b^n [f^{2n}(0)+f^{2n}(1)], \qquad (5)$$

and the right-hand side is an integer.

As both $\sin(\pi x)$ and $x(1-x)$ lie between 0 and 1 we have

$$0 < \pi a^n I < \pi a^n \int_0^1 \frac{dx}{n!} = \frac{\pi a^n}{n!}.$$

If n is large enough, this quantity is less than 1, so the left-hand side of (5) cannot be an integer. The previous result is contradicted, and we must abandon the assumption that π^2 is rational.

This proof is due to Niven; he also uses similar methods to show that various trigonometric quantities are irrational (Niven[3]).

5. *Buffon's needle problem*

We conclude with an account of a statistical method of measuring π. This was devised by Buffon, a naturalist associated with investigations into the geometry of the bee's cell. A needle of length l is dropped at random on to a plane ruled with parallel lines at unit intervals. If $l < 1$, the probability that the needle cuts at least one line is $2l/\pi$; otherwise it is

$$\frac{2l}{\pi}(1 - \cos \phi_0) + \left(1 - \frac{2\phi_0}{\pi}\right), \quad \text{where} \quad \sin \phi_0 = \frac{1}{l}. \qquad (6)$$

Repeated experiments with a count of favourable cases leads to a Monte Carlo determination of π. Kahan[5] gives an interesting account of a carefully designed experiment. This includes a modified procedure designed to reduce the effect of random variations on the answer. Thus we can replace the parallel grid by a square grid with (b) suitably modified. This divides the effect of random errors by nearly 4, and further improvements are possible.

Let the centre of the needle be at distance x from the nearest line, and its direction make an acute angle ϕ with it, as in Figure 5(*a*). With $l < 1$, Figure 5(*b*) illustrates the rectangle over which (x, ϕ) ranges; we assume all points in this area occur with equal probability. Points leading to an intersection have $2x$ less than $l \sin \phi$, corresponding to the shaded area. The probability of an intersection is the area ratio

$$\int_0^{\frac{1}{2}\pi} \int_0^{\frac{1}{2}(l\sin\phi)} dx \, d\phi \Big/ \frac{\pi}{4} = \frac{2l}{\pi}.$$

If l exceeds unity sketch (c) shows that the probability is

$$\frac{4}{\pi}\int_0^{\phi_0}\int_0^{\frac{1}{2}(l\sin\phi)} dx\,d\phi + \frac{4}{\pi}(\tfrac{1}{2}\pi - \phi_0)\cdot\tfrac{1}{2},$$

which reduces to (6) above.

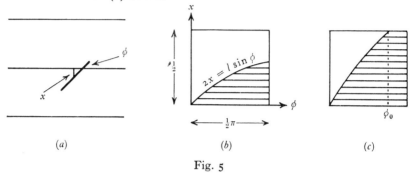

(a) (b) (c)

Fig. 5

This is a classical result in the theory of geometrical probability. Kendall and Moran[1] give an interesting account of this relatively neglected field. We shall use the above result for a needle of length l to deduce the probability that a convex curve of maximum breadth less than or equal to unity will intersect a line of the parallel set.

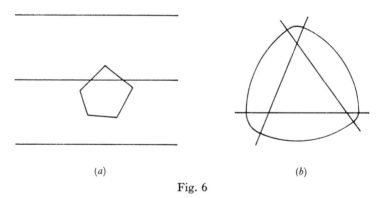

(a) (b)

Fig. 6

In Figure 6(a) we illustrate a convex polygon of n sides and of maximum breadth less than or equal to 1. The lengths of these sides are l_1 to l_n. We assume that the probability of cutting side j is $p(j)$, while the probability of cutting both j and k is $p(j, k)$. If side j is cut, then one other side is also cut, so that

$$p(j) = \sum_{(k)}' p(j, k), \tag{7}$$

162

the dash denoting the absence of $p(j, j)$ in the sum. If the polygon intersects a line then just two sides are cut, so the probability P of this happening will be

$$P = \Sigma\Sigma'p(j, k),$$

where the sum is taken over all possible different pairs j, k. If we sum (7) over all sides we get this latter sum counted twice, for $p(j, k) = p(k, j)$. By the result proved earlier, $p(j) = 2l_j/\pi$, so that

$$P = \tfrac{1}{2}\Sigma p(j) = \tfrac{1}{2}\Sigma \frac{2l_j}{\pi} = \frac{C}{\pi},$$

where C is the polygon's circumference. By replacing a convex curve by an approximating polygon, and proceeding to the limit, the result is seen to hold for curves too.

If the curve is of breadth 1 in all directions intersection is certain and P is unity. Moreover, by the above formula, $C = \pi$ for all such curves, a result due to Barbier that we shall meet again in the next chapter. Figure 6(b) is made up of 6 circular arcs with centres at the 3 vertices of the triangle. It is of constant breadth, and provides a direct confirmation of Barbier's theorem, since we see readily that its circumference is π times the constant breadth.

References

1. M. G. Kendall and P. A. P. Moran. *Geometrical Probability* (Griffin, 1963).
2. F. Klein. *Famous Problems of Elementary Geometry* (Dover, 1956).
3. I. Niven. *Irrational Numbers*, Carus Monograph no. 11 (Wiley, 1956).
4. G. Polya. *Mathematics and Plausible Reasoning*, vol. 1 (Princeton, 1954).
5. B. C. Kahan. 'A practical demonstration of a needle experiment designed to give a number of concurrent estimates of π.' *J. R. Statist. Soc.* (A), vol. 124, part 2, p. 227 (1961).

15

Rotors and curves of constant breadth

1. *Curves of constant breadth*

The distance between parallel tangents to a circle is constant, but it is not the only curve with this property. We consider a closed convex curve and a pair of parallel support lines (see Figure 3 of Chapter 13). The

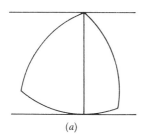

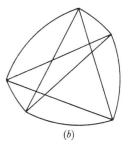

(a)   (b)

Fig. 1

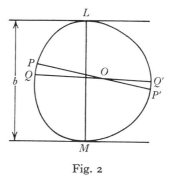

Fig. 2

distance between them is the curve's breadth in the direction of these lines. We shall investigate curves for which the breadth is the same in any direction, and start with some simple examples.

In Figure 1 (*a*) the curve is defined by 3 circular arcs with centres at the vertices of an equilateral triangle whose side equals their radii. This is the Reuleaux triangle; in Figure 1 (b) it is generalised by starting with a crossed pentagon, not necessarily regular, but with sides of equal length. It can be expanded by a constant amount to remove discontinuities in slope; the result for the Reuleaux triangle is seen in Figure 3(b). Figure 6(b) of Chapter 14 shows how a 6-arc curve of constant breadth can be formed from any triangle. By using a pentagon a 10-arc curve can be constructed, and so on.

So far the curves discussed have been composed of circular arcs, but we shall now find a far more general procedure. There cannot be two points inside or on a curve of constant breadth b further apart than b. If there were, we could draw lines through them perpendicular to their join, and the curve's breadth in this direction would exceed b. The line LM joining points of contact of parallel support lines must therefore be perpendicular to them both, for LM cannot exceed b.

If we take P and Q close together and draw the normals there to meet the curve again at P' and Q', the lines PP', QQ' will cut at O, the centre of curvature at both P and P'. If the arc LPM is arbitrary, but has the property that all circles of radius b touching it enclose the curve, we can complete a curve of constant breadth as follows. Along the normal at P measure distance b to find P', then the locus of P' is the other part of our curve. By our assumption about tangent circles, the centre of curvature at P lies inside PP', and the resulting curve is convex. This heuristic proof can be supplemented by Rademacher and Toeplitz's[3] rigorous discussion.

We now consider the circumference C of our curve. Let angle POQ be $\delta\theta$, then we have

$$PQ \simeq OP.\delta\theta, \quad P'Q' \simeq OP'.\delta\theta, \quad C = \int_0^\pi (OP+OP')d\theta = b\pi.$$

This surprising result is due to Barbier; Lyusternik[2] gives a rigorous proof. A proof based on geometrical probability was given at the end of Chapter 14. A direct verification is easy for the curves composed of circular arcs discussed above. We shall prove analogous results by a quite different method later.

It was proved in Chapter 13 that any convex curve possesses a circumscribing equiangular hexagon with central symmetry. If the curve is of constant breadth this hexagon is regular, a fact Eggleston[1] uses in proving the Blaschke–Lesbegue theorem. This states that, of all curves of given constant breadth b, the Reuleaux triangle has the smallest area. The curve of largest area is the circle; this follows from the fact that its circumference is determined by b. Among all curves of given circumference the circle has greatest area, and it is also a curve of constant breadth.

2. *Mechanical applications*

These curves have a number of mechanical applications. In Figure 3 (a) we enclose the curvilinear triangle in a square, and it can be rotated so as to maintain contact with the sides of the square. Curves of constant breadth are rotors for a square; we discuss equilateral triangle or Δ-rotors later. This property led to the invention of the Watts drill for

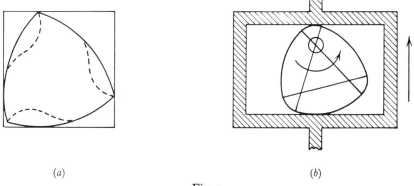

(a)   (b)

Fig. 3

square holes. A cylinder whose cross-section is a Reuleaux triangle has cutting edges formed on it by removing the regions indicated by dots in the figure, and the end is pointed. If rotated inside a square template and fed forward it drills a square hole, except for small rounded areas at each corner. The centre of rotation is changing throughout the motion, so a floating chuck is needed to impart the rotary movement. For mechanical reasons it is not feasible to use a similar technique for equilateral triangles; but, by using appropriately shaped rotors, drills for holes of 5, 6 and 8 sides have been made.

Figure 3 (b) illustrates a cam of constant breadth, in this case a modified Reuleaux triangle, rotating about one of its three original vertices. The follower, shown shaded, is stationary over two arcs each of 60°; in the diagram it is on one of these portions, and a small further rotation in the direction shown will cause it to rise. Such periods

166

of halted motion are needed in the feed mechanism for projectors of cine-film. By a suitable choice of cam shape a wide variety of motions can be generated. One of the mechanical advantages of such cams is that the drive on the follower is a positive one in both directions, in contrast to the spring loaded plunger needed with ordinary cams.

The rotating piston of the Wankel non-reciprocating engine is another example of a rotor, this time within a region of relatively complicated shape.

3. *Solids of constant breadth*

The simplest solid of constant breadth is obtained by rotating a Reuleaux triangle about an axis of symmetry, as shown in Figure 4(a).

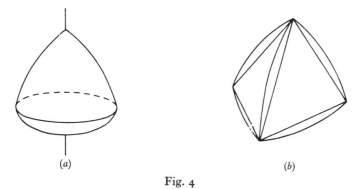

(a)   (b)

Fig. 4

We shall obtain two other such solids by a different generalisation of the Reuleaux triangle. Figure 4(b) illustrates the solid formed by constructing four spherical caps of radii equal to the sides of a regular tetrahedron, and with centres at its vertices. In the diagram it is assumed that the tetrahedron is opaque, while the spherical caps are transparent. This is not a solid of constant breadth unless suitably modified. It occurs in nature in the skeleton of the nassellarian, a tiny sea creature about 0·1 mm. in diameter. D'Arcy Thompson[5] has a beautiful drawing of it. It has, besides the curvilinear tetrahedron, six planes going through its centre and edges. Three of these, together with one of the four caps, can be seen in Figure 5(a). The angles between plane faces and between planes and spherical caps are all 120°.

In order to form the two solids of constant breadth shown in Figure 6, three edges of Figure 4(b) have to be rounded off. In the first solid these

167

edges meet at an apex, in the second they belong to one face. The new spindle-like surfaces are formed by rotating a circular arc about its chord through 60°.

In Figure 5 (b) the spherical cap opposite A cuts the plane face ACD in the lower arc CD. The cap opposite B cuts BCD in the upper arc CD. Rotating one into the other generates the required surface opposite the

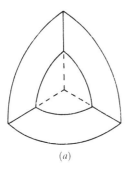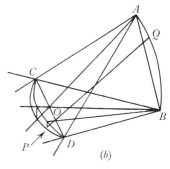

(a) (b)

Fig. 5

unmodified arc AB. There are two possible types of contact of the solid with parallel support planes. In the first, one passes through an apex, the other touches the corresponding cap at a distance equal to the side

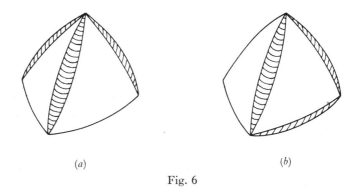

(a) (b)

Fig. 6

of the tetrahedron. In the second, one plane touches the spindle-like surface at P. The normal there is POQ, cutting the opposite arc at Q, the point of contact of the other plane. If we rotate POQ about O so that Q moves into B, it is easy to see that the line in its new position equals BD in length.

Much less is known about solids of constant breadth than about the

168

curves discussed above. There is no direct analogue of Barbier's theorem. However, Minkowski pointed out that their shadows by orthogonal projection are of constant circumference.

4. *The addition of curves*

We now describe an addition operation that provides a powerful tool for questions concerned with rotors. In Figure 7(a), we add A_1 of region C_1 to A_2 of C_2 to get the sum point (A_1+A_2). The process is one of vector addition with a fixed origin O. The sum total of all points like (A_1+A_2) is a new region (C_1+C_2), the sum of C_1 and C_2. This sum is unaltered, except for a parallel translation, if either C_1 or C_2 is moved parallel to itself. In Figure 7 (b), the arrowed vector indicates how A_2 moves into A_2'. The sum point (A_1+A_2) moves into (A_1+A_2') by the

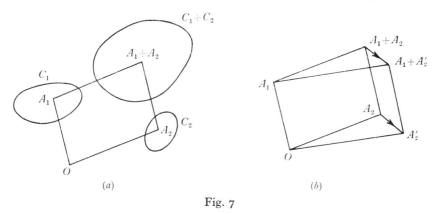

Fig. 7

same vector. Hence (C_1+C_2) also gives (C_1+C_2') by the same translation. A change of origin is equivalent to the same parallel shift applied to C_1 and C_2, and, by the previous argument, produces no alteration in the form of (C_1+C_2). Thus we may select O at any convenient point.

In Figure 8(a) we add non-parallel lines l_1 and l_2, taking O at one end of l_1. Adding l_1 to point P of l_2 is equivalent to moving l_1 until O lies on P, and it assumes the position PQ. As P takes all possible positions on l_2, PQ sweeps out the parallelogram (l_1+l_2). In Figure 8 (b) the lines are parallel, and the sum is a line parallel to both, and with length equal to the sum of their lengths.

We now prove that, if C_1 and C_2 are convex, so is their sum. Take P, Q any two points of the sum region, then P is the sum of points P_1 and P_2

in C_1 and C_2 respectively. The same is true of Q, Q_1 and Q_2. As the base curves are convex all points of P_1Q_1 lie in C_1 and of P_2Q_2 in C_2. Thus all points of the sum of P_1Q_1 and P_2Q_2 lie in (C_1+C_2). This sum is a parallelogram with diagonal PQ, so all points of PQ lie in (C_1+C_2), i.e. it is a convex region.

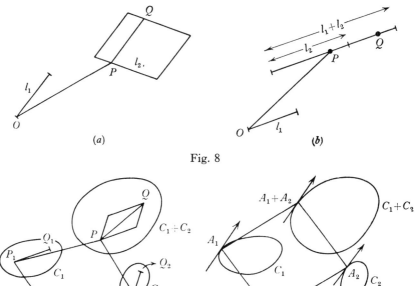

(a)   (b)

Fig. 8

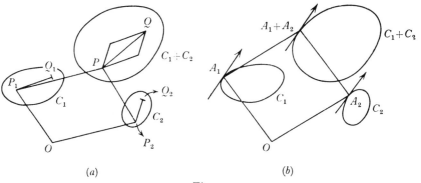

(a)   (b)

Fig. 9

Next we take A_1 and A_2 on C_1 and C_2 so that the corresponding support lines are parallel, and directed in the same sense as in Figure 9(b). All points of C_1 and C_2 lie to the right of the corresponding support lines. These support lines add up to give a parallel line through (A_1+A_2). As all points of (C_1+C_2) lie on its right and (A_1+A_2) lies on it, it is the support line there, thus parallel support lines are additive.

In Figure 10 we take O on C_1. Adding C_1

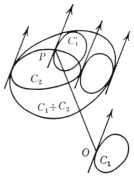

Fig. 10

to point P of C_2 is equivalent to a parallel translation that moves O on to P.

170

Thus $(C_1 + C_2)$ is generated as the envelope of C_1' when P moves around C_2. In two positions of P, support lines of C_2 are parallel to the support line of C_1 at O. Thus we see that the breadth of $(C_1 + C_2)$ in this direction is the sum of the breadths of C_1 and C_2. As O can take any position on C_1, it follows that breadths in any direction are additive. Thus curves of constant breadth add to give a new curve with the same property. When adding a circle it is convenient to take O at its centre. We see that the curve of Figure 3(b) is obtained by moving the centre of a circle round a Reuleaux triangle, an instance of the additive property just proved.

5. *The additive property of circumferences*

We now prove the remarkable theorem that circumferences are also additive. In doing so we derive another result of considerable use in the sequel.

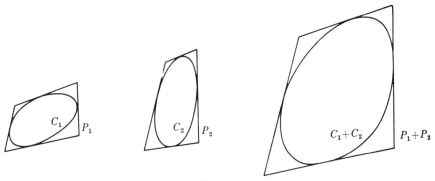

Fig. 11

In Figure 11 we have curves C_1 and C_2 inscribed in polygons P_1 and P_2 with corresponding sides parallel. The sum of these polygons is another with sides parallel to both. The length of a side of this new polygon is the sum of the corresponding sides in P_1 and P_2; moreover, the sum $(C_1 + C_2)$ is inscribed in it. These results all follow at once from the properties deduced above, and they are valuable in their own right. We next assume that the number of sides increases so that perimeters of polygons approximate to circumferences of inscribed curves. Since perimeters of polygons are additive, we see that the circumference of $(C_1 + C_2)$ is the sum of the circumferences of C_1 and C_2.

171

6. *Rotors in the equilateral triangle*

We shall describe such triangles as Δ-triangles for short in what follows. In Figure 12, *ABC* is a Δ-triangle and *AO* is parallel to *BC*, the arc *LM* is drawn with *O* as centre to touch *BC*. Its mirror image in *LM* completes a 'Δ-biangle' shown shaded and with 60° angles at its vertices.

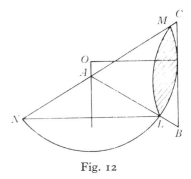

Fig. 12

Draw *LN* perpendicular to *OA*, meeting *CA* in *N*. Then the circle, centre *O*, goes through *N*, and angle *LOM* is twice angle *LNM*. This latter angle is 30° so that *LOM* is 60°, and *LM* is of fixed length. Thus the biangle will turn with arc *LM* touching *BC*. Moreover, it is easily seen that the two arcs at *M* include 60° between their tangents. So when the biangle is rotated until *M* lies at *C*, its arcs touch *AC*, *BC* there. Further rotation brings an arc in contact with *CA*, and *M* on to *BC*. In short, we can rotate the Δ-biangle inside *ABC*.

It can be proved that the perpendiculars to *BA*, *CA* at *L* and *M* meet on the perpendicular from *O* to *BC*. Their common point is the biangle's instantaneous centre of rotation.

A simpler rotor is the Δ-triangle's incircle. It can be proved that, for a given triangle, this is the rotor with greatest area, while the Δ-biangle has the least area. In Figure 16(*a*) four circular arcs of radius *d* are drawn with centres at the vertices of a square of side *d*. The curvilinear quadrilateral so formed is another Δ-rotor; Yaglom and Boltyanskii[4] give a proof.

In Figure 13 we add a Δ-biangle to a circle to produce a new convex figure. In position *B* the biangle touches the sum-curve, in *B'* one of its vertices lies on a circular arc of the curve. If we draw parallel Δ-triangles around *B* and *C*, their sum is another Δ-triangle circumscribing (*B* + *C*), thus Δ-rotors are additive, and (*B* + *C*) is a new Δ-rotor. This property is analogous to the similar one for curves of constant breadth, i.e. rotors in a square.

There is also an analogue of Barbier's theorem; all Δ-rotors in a given Δ-triangle are of the same circumference.

172

In Figure 14 we start with the Δ-triangle T_1 containing a rotor. This configuration is rotated twice through 120° about a point O on one of its axes of symmetry, giving three Δ-triangles in all. The three triangles add up to a new Δ-triangle; the three rotors give a rotor inscribed in it. The sum figures have been enlarged in the diagram for the sake of clarity. We shall prove that the new large rotor is a circle. It then follows

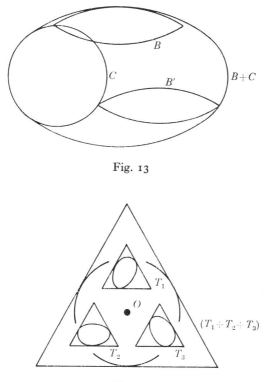

Fig. 13

Fig. 14

that the circumference of each of the three small rotors is one-third of the circumference of this circle, and so depends only on the size of the Δ-triangle T_1.

The sum-rotor is unaltered by a rotation of 120° about O, for this leaves the constituent figures unchanged. Any Δ-triangle circumscribed about it is also invariant under this rotation. For it is the sum of three Δ-triangles that are merely interchanged by the rotation. Therefore its sides must be equidistant from O. As this is true of all circumscribing triangles, and they are of the same size, the sum-rotor is a circle with

173

centre at O. The Barbier theorem for curves of constant breadth can be proved in the same way. Such curves are rotors in a square and four rotations, each of a right angle, lead to a similar proof.

7. *Some generalizations*

The Δ-rotor theory extends in a straightforward way to other regular polygons. However, of more interest is the possibility of rotors, other than circles, in non-regular polygons. A remarkable theorem, due to Fujiwara, states that, for $n \neq 4$, the necessary and sufficient conditions for the existence of such rotors are that all angles of the polygon are rational multiples of π and that it possesses an in-circle. For $n = 4$ the first condition is not necessary, thus all curves of constant breadth b constitute rotors for any rhombus with parallel sides a distance b apart.

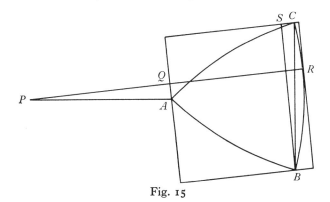

Fig. 15

Figure 15 leads to another quite different generalization. Here ABC is equilateral and $PA = AB$, the circular arc BC has P as centre, and so on for arcs CA and AB. Any rectangle is drawn circumscribing the curvilinear triangle ABC. If PQR is drawn perpendicular to two of its sides, and to BS, then angles APQ, SBC lie between 2 pairs of perpendicular lines, and so are equal. As triangles APQ, SBC are also right-angled, and $BC = PA$, they must be congruent. Thus $PQ = BS$, and the fixed length PR is the sum of sides of the circumscribing rectangle. It follows that all these rectangles have a fixed perimeter.

In Figure 16(b) the 90° biangle is another curve with circumscribing rectangles of fixed perimeter; Yaglom and Boltyanskii[4] give a proof. The addition of two curves C_1 and C_2, each with circumscribing rectangles

174

of fixed perimeter, is a curve $(C_1 + C_2)$ with the same property. To see this we consider parallel rectangles R_1 and R_2 circumscribing C_1 and C_2 respectively. The sum of R_1 and R_2 is a rectangle circumscribing $(C_1 + C_2)$, and its perimeter is the sum of the perimeters of R_1 and R_2. This sum is constant as R_1 and R_2 rotate with their sides remaining parallel.

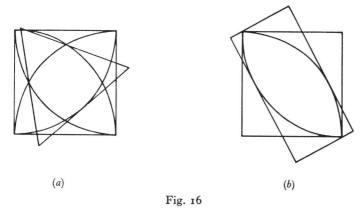

(a)   (b)

Fig. 16

Finally, these curves have circumferences depending only on the fixed perimeter of circumscribing rectangles, i.e. a generalized Barbier's theorem again applies. The proof follows the lines of that for Δ-rotors, depending on four rotations of $\frac{1}{2}\pi$, followed by addition of four equal figures.

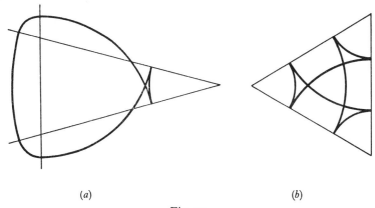

(a)   (b)

Fig. 17

So far all the curves considered have been convex. If the curvature condition on the arbitrarily specified arc LPM used in the construction of Section 1 is violated, the result will be a non-convex curve. It is

175

simpler to construct an example using circular arcs, and two are shown in Figure 17. The distance between parallel tangents is constant, so in this sense they are still curves of constant breadth, although having double points and cusps. Such a curve satisfies Barbier's theorem provided certain arcs have their lengths counted as negative in the 'circumference' of the curve.

References

1. H. G. Eggleston. *Convexity*, Cambridge Tract no. 47 (Cambridge, 1963).
2. L. A. Lyusternik. *Convex Figures and Polyhedra* (Dover, 1963).
3. H. Rademacher and O. Toeplitz. *The Enjoyment of Mathematics* (Princeton, 1957).
4. I. M. Yaglom and V. G. Boltyanskii. *Convex Figures* (Holt, Rinehart and Winston, 1961).
5. D'Arcy W. Thompson. *On Growth and Form*, abridged edition (Bonner) (Cambridge, 1961).
6. M. Goldberg. *Rotors in Polygons and Polyhedra* (Maths of Comp., **14**, No. 71 (1960).

Index

179